Advances in 21st Century Human Settlements

Indexed by SCOPUS

This Series focuses on the entire spectrum of human settlements—from rural to urban, in different regions of the world, with questions such as: What factors cause and guide the process of change in human settlements from rural to urban in character, from hamlets and villages to towns, cities and megacities? Is this process different across time and space, how and why? Is there a future for rural life? Is it possible or not to have industrial development in rural settlements, and how? Why does 'urban shrinkage' occur? Are the rural areas urbanizing or is that urban areas are undergoing 'ruralisation' (in form of underserviced slums)? What are the challenges faced by 'mega urban regions', and how they can be/are being addressed? What drives economic dynamism in human settlements? Is the urban-based economic growth paradigm the only answer to the quest for sustainable development, or is there an urgent need to balance between economic growth on one hand and ecosystem restoration and conservation on the other—for the future sustainability of human habitats? How and what new technology is helping to achieve sustainable development in human settlements? What sort of changes in the current planning, management and governance of human settlements are needed to face the changing environment including the climate and increasing disaster risks? What is the uniqueness of the new 'socio-cultural spaces' that emerge in human settlements, and how they change over time? As rural settlements become urban, are the new 'urban spaces' resulting in the loss of rural life and 'socio-cultural spaces'? What is leading the preservation of rural 'socio-cultural spaces' within the urbanizing world, and how? What is the emerging nature of the rural-urban interface, and what factors influence it? What are the emerging perspectives that help understand the human-environment-culture complex through the study of human settlements and the related ecosystems, and how do they transform our understanding of cultural landscapes and 'waterscapes' in the 21st Century? What else is and/or likely to be new vis-à-vis human settlements—now and in the future? The Series, therefore, welcomes contributions with fresh cognitive perspectives to understand the new and emerging realities of the 21st Century human settlements. Such perspectives will include a multidisciplinary analysis, constituting of the demographic, spatio-economic, environmental, technological, and planning, management and governance lenses.

If you are interested in submitting a proposal for this series, please contact the Series Editor, or the Publishing Editor:

Bharat Dahiya (bharatdahiya@gmail.com) or

Loyola D'Silva (loyola.dsilva@springer.com)

Vijayalaxmi J.
Editor

Conservation of Built Heritage in India

Heritage Mapping and Spatializing Values

Springer

Editor
Vijayalaxmi J.
Department of Architecture
School of Planning and Architecture
Vijayawada, India

ISSN 2198-2546 ISSN 2198-2554 (electronic)
Advances in 21st Century Human Settlements
ISBN 978-981-99-8334-6 ISBN 978-981-99-8335-3 (eBook)
https://doi.org/10.1007/978-981-99-8335-3

This Springer imprint is published by the registered company Springer Nature Singapore Pte Ltd.
The registered company address is: 152 Beach Road, #21-01/04 Gateway East, Singapore 189721, Singapore

Paper in this product is recyclable.

Dedicated to all my students

Series Editor's Foreword

The present volume, *Conservation of Built Heritage in India—Heritage Mapping and Spatializing Values*, edited by Dr. Vijayalaxmi Kasinath Iyer, who is based at the School of Planning and Architecture, Vijayawada in Andhra Pradesh, India, is a welcome addition to the Scopus-indexed Springer Nature book series, *Advances in 21st Century Human Settlements*. In its scope, this book series focuses on "the entire spectrum of human settlements—from rural to urban, in different regions of the world" (Springer Nature 2023). Established in January 2014, the book series features over 45 monographs and edited volumes.

Around the world, there is growing awareness of the multifarious importance and contribution of cultural heritage, including built heritage, and the need for its conservation. Under *the 2030 Agenda for Sustainable Development* (United Nations 2015), Sustainable Development Goal 11 (SDG 11), Target 11.4 aims to "strengthen efforts to protect and safeguard the world's cultural and natural heritage" (United Nations 2023). It is these buildings of precious heritage value and the places of (pre)historic importance that provide the tangible foundation for the sustenance of cultures and civilizations in the twenty-first century. Moreover, they enrich our cultural experience as well as teach us about the marvelous art and architecture of the bygone ages.

India's multifaceted architectural heritage endures despite numerous foreign invasions, rapid urbanization, industrialization, environmental pollution, natural disasters and climate change. This rich cultural heritage carries the impressive imprints of many empires, the art and architecture of the past royalty and people, and the inimitable work of craftsmanship and artisanship. Belonging to one of the oldest civilizations that exist today, the unique cultural heritage of India ranges from individual heritage buildings to entire human settlements and cities. All of this underlines the critical importance of the proper study, research and documentation, as well as the preservation and conservation of this precious cultural heritage of India.

In this larger context, this edited book is an attempt to document the current state of several buildings and city spaces of historic importance and heritage value in India. This work has been accomplished through on-location field work and follow-up analysis and documentation by conservation architects. Based on this, recommendations are made for the conservation of the specific cultural heritage under study by highlighting its relevance and importance. In putting together this edited volume, Dr. Vijayalaxmi Kasinath Iyer mentored the architectural conservationists in order for them to present their work in the form of book chapters based on their field studies.

This book will be beneficial to policy makers, architectural conservationists, architects, urban planners, urban designers, urban geographers, cultural heritage historians, researchers and postgraduate students interested in cultural heritage buildings and city spaces of rich heritage value.

For their contribution to the study of cultural heritage in India, Dr. Vijayalaxmi Kasinath Iyer and the contributing authors are worthy of our commendation.

Bharat Dahiya
Director
Research Center for Sustainable Development and Innovation
School of Global Studies
Thammasat University
Bangkok, Thailand

Extraordinary Professor
School of Public Leadership
Stellenbosch University
Western Cape, South Africa

References

Springer Nature (2023) *Advances in 21st Century Human Settlements: About this book series*. Springer Nature, Singapore. See: https://www.springer.com/series/13196

United Nations (2015b) Transforming our world: the 2030 agenda for sustainable development, A/RES/70/1. https://sustainabledevelopment.un.org/post2015/transformingourworld/publication. Accessed 30 Oct 2023

United Nations (2023) Sustainable development goal 11: make cities and human settlements inclusive, safe, resilient and sustainable. https://sustainabledevelopment.un.org/sdg11. Accessed 30 Oct 2023

Preface

The book Conservation of Built Heritage in India is an outcome of field investigations in various domains and aspects of Architectural Conservation by Conservation Architects and Sustainable Architects. The work is carried out as part of a course 'Research Methodology' handled by me at the School of Planning and Architecture, Vijayawada, India. This book contains chapters on the following areas of Conservation: 1. Historic districts of a city/town, 2. Large mansion, fort and temples, reminiscent of their rich culture, but losing their function with changing context, 3. Public perception and acceptance of built heritage and adopting vernacular solutions to design and 4. Assessing the built heritage for its potential for adaptive reuse as a protective mechanism.

Challenges of Built Heritage in India arises from the complexity of heritage context combined with developmental pressures. Priorities and legislative frameworks also pose a challenge. There are many valuable heritage structures which are unrecognized and therefore vulnerable to perishing in the process of development. The first and second parts of the book deal with such cases. The Padmanabhapuram Town is known largely for its Palace and Temple. The historic significance of the town and its tourist potential is still lesser know. The Chikkajala Fort in Bengaluru is another classic example of how a historic structure gradually succumbed to development. These are covered in Chaps. 1 and 2. The second part of the book covers the conservation of a lesser known buildings Chap. (3—Wadas of Nagpur), fort (Chap. 4—Gwalior Fort) and temple styles (Chap. 5—Temples of Nagpur). India has unlisted and unprotected heritage, which could perish due to a lack of documentation and awareness. This section is an initiative to highlight their cultural richness.

In order to sustain Built Heritage, it is important to know public perception and desire for individual or community-inclusive initiatives. Chapter 6 deals with assessing the mood of adopting vernacular solutions in urban single residences. It assesses the desire of the public to be rooted in their culture through design solutions despite the changing context. Chapter 7 studies cultural heritage conservation to promote community well-being. Studies of awareness and the importance of local cultural heritage and conservation for promoting community well-being form part of this chapter. Once the built heritage worthy of conservation is identified, the technique of conservation becomes important. Chapters 8 and 9 deal with assessing buildings of heritage value with case study examples for their Adaptive Reuse Potential through the ARP model.

Considering the vastness of the unprotected built heritage, attempts at documentation of these structures, assessing public acceptance and assessing the potential for the structures to be reused will help to conserve the buildings. This book is one such attempt, carried out through case studies from India, in order to highlight the need for conserving the rich history through research, education, advocacy and practice. The historic significance and the condition mapping of the buildings are carried out through primary field studies by the authors who were students of the School of Planning and Architecture, Vijayawada, India. This book serves as a rich database of literature as well as conservation strategies for heritage conservationist, students of architectural conservation and researchers.

I express my sincere thanks to Dr. Bharat Dahiya, Dr. Loyola D'Silva, Ms. Shalini Monica and Ms. Diya Ma, Ms. Niyati and Dr. Ramesh Srikonda for their great support during the entire process in the preparation of the final book.

Dr. Vijayalaxmi J.
School of Planning and Architecture
Vijayawada
An Institute of National Importance Under the Ministry of Education
Government of India

Contents

About the Editor

Dr. Vijayalaxmi J. has completed her B.Arch. from School of Architecture and Planning, Anna University in First Class with Distinction. She went on to pursue the M.Arch. Programme from School of Architecture and Planning, Anna University in 1997. She completed her Ph.D. as an AICTE QIP Scholar in the year 2008 in the topic of 'Thermal performance of naturally ventilated residential buildings with varying opening sizes and orientation in hot-humid climate- a case of Chennai City' from Anna University, Chennai. Subsequently, she was awarded the Erasmus Mundus Fellowship for Academic Profile, and completed her Higher Research in Politechnico di Milano in 2010 in the area of 'Smart Materials and Nano technology in the Construction Industry'.

She has over 25 years of experience in teaching architecture and specializes in Sustainable Built Environment and Climate Responsive Architecture. She is currently the Professor and Head of the Department—Architecture at the School of Planning and Architecture, Vijayawada, An Institute of National Importance under the Ministry of Education, Government of India and was also the Dean (Research) there. She has completed funded research projects sponsored by the Government of India and has several awards to her credit including the 'Yuva Ratna Award', 'Best Teacher Award' and the 'Faculty Lead Award of Merit for the Solar Decathlon India'.

She has published many technical papers in peer-reviewed blind refereed International journals and Scopus indexed books and has presented technical papers in many International Conferences including at the Harvard, USA, Politechnico di Milano, Italy, NRCC, Canada, Aristotle University of Thessaloniki, Greece sponsored Conference.

She has travelled widely in her academic quest to countries such as Greece, France, Canada, Singapore, Malaysia, The United States of America and Italy. She has several publications in International peer refereed Journals and Edited Chapters in reputed Books. She has also Authored a Book titled' Building Thermal Performance and Sustainability' as part of the Book Series 'Lecture Notes in Civil Engineering'. She has also edited a Scopus indexed book titled 'Building Construction and Technology' as part of the Book Series 'Lecture Notes in Civil Engineering'.

Conservation of Heritage in India-Challenges, Issues, Agencies, Technological Advancements

Vijayalaxmi J.

Abstract Conservation of heritage is important as it entails protecting the cultural resources that define the very essence and character of a civilization. The various aspects of heritage which needs to be conserved are the arts and crafts of a place, the buildings of heritage importance and the cities, towns and places of heritage value. Arts and crafts include traditional textiles (handloom weaves and patterns and the method of weaving), Carpets and embroidery making, Handicrafts made with vernacular materials such as bamboo, cane, grass, wood. Paintings and murals such as Warli and Madhubani on fabric as well as on Canvas, Brassware and Earthenware along with various finishes, Unique jewellery such as Minakairi, Chettinadu jewellery and leather products tell tales of historic times. It is easier to preserve these products, but the significance lies in conserving the process too. It is utmost difficult to conserve buildings and cities of historic importance due to the urbanization and also the cost involved. Architectural conservation is the preservation of important buildings or architectural values. Ever since it first appeared, the preservation and repair of architecture has been seen as an essential virtue. With the intention of safeguarding the heritage so that it has an embodied reality for its interpretation, efforts have been made to understand and interpret it in order to show how the world was before us. This Chapter deals with the various bodies in India that work towards preservation and conservation. Efforts at conserving various religious art, folk art and commercial art by various Governmental bodies, NGOs and private players have been assessed. The conservation strategies have been discussed too.

Keywords Conservation of Indian arts · Crafts · Adaptive reuse · Government bodies for conservation · Private initiatives for conservation · Indian National Trust for Art and Cultural Heritage (INTACH) · National Handicrafts Development Programme (NHDP) & Comprehensive Handicrafts Cluster Development Scheme (CHCDS) · Sustainable Development Goal (SDG 11)

Vijayalaxmi J. (✉)
School of Planning and Architecture, Vijayawada, India
e-mail: vijayalaxmij@spav.ac.in

Vijayalaxmi J. (ed.), *Conservation of Built Heritage in India*, Advances in 21st Century Human Settlements, https://doi.org/10.1007/978-981-99-8335-3_1

1 Conservation Scenario in India

It is important to conserve heritage, as it has 'value' for various reasons (Taylor and Cassar 2008). What is heritage is in itself contextual to a place. India is rich in its tangible heritage of arts, crafts (including jewellery, textiles) and architecture (including monuments). Historical buildings, monuments, and archaeological sites comprise immovable heritage (Tolou Del et al. 2020). The need to safeguard historic monuments and hand it over to the next future generations is well recognized (Petzet 2004). The protection of historic monuments has been crucial to India's nation-building and statehood process. Until the middle of the twentieth century, India was a colony of the British Empire. As complex as India's history of nationalism and state formation are the politics of history, legacy, and the preservation of historical structures. India is a nation that is sometimes referred to as a subcontinent. It is made up of numerous language, regions, ethnic and religious groupings, as well as profound social and cultural disparities that are ingrained in the caste system. Furthermore, it is challenging to understand heritage and preservation in modern-day India due to its colonial past. The first attempt at conserving was the passing of The Ancient Monuments and Archaeological Sites and Remains Act, 1958 aimed at preserving ancient monuments and sites of archaeological importance in India. However, it may be noted, that the act provides for preservation as the technique for conservation. The National Heritage Sites and Regulation Authority Act was enacted in 2008in order to have a complete proposal to protect heritage sites. This included preserving, conserving and developing the national heritage sites, as preservation alone could not be considered as a viable technique of conservation. The quantum of historic sites in India is humungous and hence the entire body of work need not be handled by the Government alone. UNHSRA collaborates with government agencies, NGOs, and local communities.

2 Various Bodies Working on Conservation in India

The main aim of these bodies, in general, is to strategize and conserve the buildings and preserve the historical accuracy and genuineness. Some of these bodies consider preserving not only tangibles but also the intangibles including social, economic, and environmental factors. Conservation efforts involve a multidisciplinary approach, combining architectural, historical, and scientific expertise. Some of the important orgnizations working toward conservation in India are

i. INTACH
ii. Tata Trusts
iii. INHCRF
iv. National Museum
v. Archaeological Survey of India.

Some of the Bodies functioning in India and Internationally are

i. The Lakshmi Mittal and Family South Asia Institute
ii. UNESCO
iii. Global Heritage Fund
iv. Aga Khan Trust for Culture.

3 Major Challenges or Issues of Conservation in India

Since India gained independence from British domination, its historical experience of colonialism and its confrontations with its colonial past have been intertwined with politics of history and heritage. In postcolonial India, the concept of cultural custodianship—which the colonial state of India fostered and which led to the bureaucratization of the business of conserving culture—was enhanced. As a result of the British Empire's heritage, which held that a strong state was required to unite a country made up of diverse communities, the bureaucratic structure of colonial authority endured. Thus, heritage and the preservation of historic buildings were included into the process of creating a state and a country. Preserved monuments were now regarded as having "national importance," The ASI's bureaucratic approach has shown that it is unable to manage this level of complexity, and its ongoing commitment to a preservation strategy that was implemented in India under quite different circumstances is evidence that it is unable to satisfy the needs of heritage-making in postcolonial India because of the large number of monuments, which are a legacy of our rich heritage. INTACH and other non-governmental heritage organizations have demonstrated a greater understanding of community rights (http://www.intach.org/, https://www.tatatrusts.org/, http://inhcrf.org/, asi).

Lack of funding, pollution and the effects of climate change on cultural objects are major problems affecting heritage conservation in India and around the world. Sea level rise (Howard et at. 2008, ICOMOS 2019, Sabbioni et al. 2006) thermoclastism (Yaldiz 2010; Cassar and Pender 2005; Camuffo 2019) potential damage due to corrosion (Bonazza et al. 2009, Fuente et al. 2013, Ciantelli et al. 2018) air pollution and acid rains (Haneef et al. 1992, Grontoft 2019, Brimblecombe and Grossi 2008). Biological degradation due to algae and flora (Cutler et al. 2013, Lefèvre 2014, Prieto et al. 2020) and fungal attack (Haugen and Mattsson 2011, Orr et al. 2018, Nagar et al. 2021) are the major climate change factors that affect the heritage structures. The Taj Mahal was impacted by pollution despite attempts being made to remedy it (Nagar et al. 2021).

India is now ranked sixth in the world for having the most UNESCO-recognized landmarks, trailing only Italy, which has the most with 58 structures.

3.1 Issues in Ensuring Safety of Heritage Structures in India

- Heritage building conservation is a multidisciplinary engineering effort, not just a concern for aesthetics and architecture, with structure safety being a key factor. In India, there are no formal mechanisms that identify the necessity of using scientific instruments for diagnostics and architecture and quantitative evaluation of residual capacity prior to deciding on a strategy for a repair or strengthening.
- A major barrier to historic preservation is the divergence between contemporary engineering education and conventional knowledge of building materials and methods.
- The contemporary post-disaster interventions in heritage structures, which are frequently improvised, ad hoc, and semi- or non-engineered, can best be described as repair. Retrofit or preventative action is preferable but necessitates a complete plan. Important aspects of conservation, such as the choice of intervention's reversibility and documentation of the intervention, are accepted as part of the process in international practice; however, national and state agencies carrying out post-disaster interventions have not yet internalized such an approach.
- India has a sizable collection of historical buildings, which needs to be addressed through a formal platform that focuses on the structural safety of those buildings. However, a major bottleneck in India is the lack of sufficient labor, both in terms of quality and quantity, to handle the enormous responsibility of comprehending and safeguarding the numerous heritage structures from natural disasters.
- For carrying out conservation engineering in India, there are currently inadequate professionals and trained employees available. All monuments of national importance must be carefully assessed scientifically and updated. According to the most recent National Conservation Policy), this poses a significant challenge due to severe lack of trained labor in the field (NPC–AMASR 2014). As a result, India faces a significant problem in expanding its competence in conservation engineering, and the current efforts are insufficient and unorganized. There is a glaring divergence between academic institutions and implementing agencies, as evidenced by the fact that even solitary efforts to conduct research on heritage structures fail to provide outcomes that are helpful and implementable (Menon 2014).

Now, let us assess the conservation measures in India pertaining to

- Indian Arts and Crafts
- Buildings of historic significance
- Towns or parts of Towns with Heritage significance.

4 Conservation of Indian Arts and Crafts: Important Forms and Attempts Made at Conservation

India has a rich cultural and historical heritage. Indian handicrafts were a significant aspect of life in the past and are today seen as even more essential because of the human resource involved in creating such exquisite works of art. Since they are the primary source of income for many communities, handicrafts are an essential component of life for Indians. This is contrary to many other western nations, where crafts are valued as luxury goods.

The art and craft forms of India can be broadly classified as belonging to one of three types as described in Fig. 1 (Types of handicraft in India 2023).

1. Religious Craft
2. Folk Craft
3. Commercial Craft.

Fig. 1 Broad classification of craft forms in India

4.1 *Research and Documentation*

- Scholars and researchers' study, document, and publish findings about traditional crafts, contributing to their preservation and significance.
- Scholars like Dr. Anamika Pathak research and document traditional crafts like Warli painting, contributing to their documentation and preservation.

4.2 *Digital Platforms and New Media*

- Embracing technology to create virtual heritage representations, animations, and interactive experiences to educate and raise awareness about traditional arts and crafts (Kolay 2016).

4.3 *Collaborations and Partnerships*

- Collaborative efforts between designers, artisans, entrepreneurs, and government bodies help blend traditional skills with contemporary designs, preserving and promoting crafts.
- For example, Fabindia: Collaborates with artisans across India to create contemporary products while retaining traditional craftsmanship, helping sustain local crafts (Dasra 2013).

5 Conservation Methods for Arts and Crafts

The Government is taking great efforts to preserve the arts and craft. The three ways in which the arts and crafts can be preserved is through

i. Transferring the knowledge to the next generation to ensure the art or craft form does not vanish.
ii. The artists must be given us recognition and a dignified opportunity.

New talent and new technologies must be integrated to ensure preservation.

There are atleast 52 art and craft forms that need preservation today. These include Aipan Art, Assamese Miniature Painting, Bani Thani, Reverse Glass Tanjore Painting, Rogan Art etc.

This can be carried out through the following means:

- Repairing and restoring damaged artifacts using compatible materials and techniques.
- Cleaning through mechanical, solvent, and steam methods to remove dirt and accretions.

- Documentation to record the history, materials, and techniques of artworks.
- Capacity building through training artisans in traditional skills, design innovation, and business practices.
- Digital documentation and virtual representations to preserve and share cultural heritage.
- Using new media like game designs and animations to educate and engage the younger generation.
- These combined efforts and methods contribute to the conservation and preservation of Indian arts and crafts, ensuring their continuation for future generations.

5.1 *Government Initiatives*

Although the Government of India, the private sector, and non-profit organisations are all participating in the sector, their responsibilities have developed in isolation, with little specialisation and a lot of overlap. However, there hasn't been a coordinated and unified effort to address the issues facing the crafts ecosystem on a policy level, with the majority of governmental offerings enmeshed in bureaucracy and inefficiencies. The Indian government has established a number of institutions, schemes, and welfare programmes for the crafts sector since independence. As industrialization rose to prominence as a top national aim in the 1960s and 1970s, the government began to consider crafts as more of a "sunset industry" and shifted its attention to welfare programmes rather than viewing them as a sector with the potential to boost India's economy. Currently, the ministry of textiles has nearly artificially divided crafts into two categories: handicrafts and handlooms, with two different development commissioners in charge of each. In addition, 16 other ministries, including the Ministry of Women and Child Development, the Ministry of Tribal Development, the Ministry of Rural Development, and the Ministry of Agro and Rural Industries, are responsible for the development of crafts.

5.2 *National Handicrafts Development Programme (NHDP) and Comprehensive Handicrafts Cluster Development Scheme (CHCDS)*

Handicrafts are vital in the unorganized economy.

- Mainly rural, found in remote and backward areas.
- Began as part-time rural work, now a thriving economic activity due to domestic and global demand.
- Offers potential growth for artisans and newcomers, generating employment and exports.

- Challenges include lack of education, capital, tech exposure, market info, and institutional support.
- Addressing challenges via National Handicrafts Development Programme (NHDP) and Comprehensive Handicrafts Cluster Development Scheme (CHCDS).
- Focus on integrated development involving artisans, empowering and ensuring sustainability.

The various services and components of NHDP and CHCDS schemes are

- Marketing Support & Services
- Skill Development in Handicraft Sector
- Ambedkar Hastshilp Vikas Yojana
- Direct Benefit to Artisans
- Infrastructure and Technology Support
- Research & Development (handicrafts.nic.in).

The majority of government funds have been directed towards specific sectors, often in collaboration with non-profit organizations, yielding notable results:

Welfare Initiatives: A significant portion of government resources has been dedicated to uplifting the underprivileged, particularly within the crafts sector. Noteworthy programs like the Rajiv Gandhi Shilpi Swasthya Bima Yojana and the Janshree Bima Yojana for Handicrafts Artisans have been established. These initiatives extend health and life insurance coverage to artisans aged 18–60 and their three dependents. Partnerships with the Life Insurance Corporation of India further fortify this social safety net (Dasra 2013).

Market Connectivity: The government's commendable efforts in the crafts sector encompass the establishment of Dilli Haat, an innovative open-air crafts market situated in the national capital. Inspired by traditional village markets, this unique platform allows artisans from across the nation to rent display spaces for a two-week duration. Collaborating with the Dastkari Haat Samiti, a non-profit organization supporting artisans nationwide, Dilli Haat has emerged as a pivotal conduit connecting artisans with consumers. Furthermore, the Development Commissioners for Handicrafts and Handlooms operate an array of emporiums across the country, serving as retail hubs for crafts originating from diverse regions.

Nurturing Capacity: In recent years, the Government has refocused its attention on promoting sustainable craftsmanship by actively involving artisans. Capacity building initiatives, under the guidance of the Development Commissioner of Handicrafts, play a pivotal role. These encompass multifaceted efforts such as honing skills, integrating technology, providing design and marketing support, and revitalizing dormant crafts. The Development Commissioner for Handlooms oversees Weavers Centers dispersed across many regions, collaborating with the National Center for Textile Design to offer market-oriented design insights to handloom weavers.

Celebrating Excellence: Annually, the Ministry of Textiles acknowledges the exemplary contributions of master craftspeople through the conferment of national awards. These coveted honors not only amplify the prestige of these artisans but also serve as a hallmark of outstanding craftsmanship, symbolizing a mark of excellence and quality endorsement.

5.3 Private Sector Initiatives

Private enterprise has served as the predominant channel for crafts production, accounting for over 95% of the total output. Broadly categorized, private engagement in the crafts sector can be classified into four primary forms:

Traditional Intermediaries: Throughout history, private involvement within the crafts sector has been predominantly led by master craftsmen, traders, and exporters. These entities have traditionally aggregated production, facilitated market connections, and extended financial support to artisans. However, it's worth noting that these dynamics haven't always been equitable, often resulting in artisans receiving disproportionately low compensation for their creations.

Mainstream Retail: A handful of successful crafts-based enterprises have emerged, markedly enhancing the livelihoods of numerous artisans while simultaneously furnishing models for future market-driven ventures. However, these success stories remain isolated and necessitate a more systematic replication approach.

Social Enterprises: A more contemporary trend, these businesses have sprouted from non-profit origins or hybrid frameworks harmonizing social and commercial objectives. This novel wave underscores a dynamic wherein businesses strive to achieve both societal impact and financial sustainability.

5.4 INTACH Conservation Institutes (ICI) and Its Initiatives

The institutes have been setup to work on the scientific investigations towards conservation and restoration of the vast repository of art objects in the country. A team of highly skilled conservators are engaged in the conservation process across INTACH Conservation Institutes in different geographies of the country. Their initiatives are in the following domains.

5.4.1 Stone Sculptures: Stone Sculptures Have Been Restored Through

1. Repairing broken sculptures with compatible materials.
2. Cleaning using mechanical, solvent, and steam methods.

Fig. 2 Conservation of flora sculpture before and after. *Source* (heritageici.intach, 2020/03/2)

3. Consolidated to prevent further deterioration.
4. Joined broken parts with dowels, filled gaps with compatible material.

The Flora Fountain, Mumba Chowk (Martyr's Square) was built in 1864 and is an exquisitely sculpted architectural heritage monument of Roman goddess Flora in the heart of South Mumbai. Over the years, the Portland stone fountain was damaged due to weather, depositions, algal growth and loss of parts (Fig. 2).

INTACH Conservation Institute team has restored the sculpture using steam machines to remove the multiple layers of deposition, removal of all previous improper unscientific cement plaster repairs and fillings (Fig. 3). Algaecide treatment is used to arrest the biological growth as well as consolidation of stone. Further, filling and grouting of joints and gaps and reintegration of losses including fixing of carved hands and other loose, detached, broken parts was undertaken (Fig. 2).

In a similar manner, the statue of Lord Surya (Sun God), which was found during flattening of a field in Gorakhpur was restored by ICI (Fig. 4), as the stature had suffered severe damage.

5.4.2 Paintings

Paintings of private collectors, museums and palaces have been restored by INTACH. The work includes:

1. Addressed damage due to humidity and temperature fluctuations.
2. Treated fungal attack, warping, rust stains, cracks, flaking paint, and joint separation.

Fig. 3 Removal of Stubborn incrustations and steam cleaning. *Source* (heritageici.intach, 2020/03/2)

Fig. 4 Conservation of Lord Surya's Statue at Sonmati Mata Temple, Gorakhpur. *Source* (heritageici.intach, 2020/03/9)

3. Cleaned mechanically and with solvents.
4. Consolidated to prevent flaking, reintegration of losses.
5. Added wooden bars for stability (Fig. 5).

Fig. 5 Mysore painting, before and after conservation, *Source* (Sita Sings the Blues)

5.4.3 Taxidermy Conservation

The various works carried out as part of taxidermy conservation include

1. Repaired leopard skin with fur shedding, damaged claws, eyes, and teeth.
2. Fumigated to eliminate insect infestation.
3. Locally consolidated fur and cleaned mechanically and with solvents.
4. Mended tears and patches with compatible material.
5. Restored broken parts, attached new lining (Fig. 6).

Fig. 6 Taxidermy restoration by INTACH

5.5 Conservation of Museum Artworks and Private Collections

In the year 2020–2021, the ICI Bangalore engaged in conservation efforts as follows.

5.5.1 Government Museum, Bangalore, and Venkatappa Art Gallery

1. Conserved 175 artworks in total.
2. Undertook a project to conserve 118 artworks (86 from Government Museum and 32 from Venkatappa Art Gallery).
3. Artworks included various media like oil on canvas, miniature paintings, traditional South Indian paintings, prints, watercolors, and line drawings.
4. Completed conservation of 21 artworks in the third phase, ongoing work on the remaining pieces.

5.5.2 The Roerich Collection Project

1. Undertook conservation of 237 artworks from the Roerich Collection at Venkatappa Art Gallery.
2. Completed conservation of 20 artworks in the first phase and 40 artworks in the second phase by August 2021.
3. Ongoing conservation efforts for 30 works in the second phase.

Significant Objects Conserved:

1. Conserved a panel painting depicting the 'Twelve feasts of the Orthodox Church' from Russia, executed in tempera technique on wood.
2. Undertook the challenging conservation of a picchwai 'Srinathji' painting on textile with water-based colors. The artwork arrived in fragile condition with friable pigment layers, particularly retaining brilliant yellow colors.
3. The institute's conservation work encompasses a wide range of artistic media and styles, reflecting their dedication to preserving and restoring significant artworks.

6 Preservation Through Virtual New-Media

Traditional Indian art and craft represent cultural heritage but face threats due to lack of awareness. Traditional art is converted into contemporary forms like game designs and animations. The younger generation gets engaged as they play a crucial role in cultural heritage preservation. This will make traditional art accessible through new-media, addressing possibilities and constraints. This will enable user-driven design to effectively transmit traditional art knowledge to the new generation (Kolay 2016).

6.1 Role of New-Media in Virtual Heritage

There are primarily two ways that new media applications can promote social awareness and the preservation of virtual heritage:

- Digital representations of cultural assets that are virtually simulated; this type of intervention belongs to the museological category.
- Visual communication design interventions that reflect cultural heritage through narrative story-telling, such as game designs and animations as shown in Fig. 7 (Kolay 2016).
- Collaboration and co-creation between visual designers and artisans are essential for the sustainability of craft guilds and cultural heritage. This partnership should encompass an educational approach that preserves the authenticity of art and craft practices while enlightening a wider audience.
- The potential of new-media applications for heritage preservation and public awareness deserves further exploration. Employing technology in new-media applications, a visual communication design can be developed to document data, depict cultural heritage, and share knowledge across diverse audiences, transcending geographical barriers. Utilizing interactive communication and gamified learning methods can significantly engage users in a co-evolutionary learning process.
- Numerous traditional Indian art forms possess the inherent capability to seamlessly integrate into contemporary new media applications. A visual exploration of these traditional arts within this modern framework is imperative to raise awareness before the invaluable wealth of our cultural heritage vanishes along with fading craft guilds (Kolay 2016). The development of goddess Durga in the game 'Wrath of Durga' is shown in Fig. 8.

Fig. 7 Ramayana as integrated in a narrative. *Source* (Sita Sings the Blue, The Stitches Speak)

Fig. 8 Development of Goddess Durga through doodles, and final design developed

7 Conservation of Individual Structures and Parts of Towns

The city's identity includes historical cores, which are concentrations of architecturally and historically significant monuments and structures. The concepts of sustainability and conservation deal with the same principle of preserving historical significance as an unrenewable resource is a piece of history forever lost upon demolition (Chandan and Kumar 2019). Cultural heritage is continually under threat globally due to a variety of pressures, which include rapid urbanization, rising housing demand, deteriorating infrastructure, and socio-cultural changes, despite being acknowledged as a focus for creating sustainable cities and communities. The establishment of methods that take into account the diversity of heritage values, as well as the careful integration of heritage urban conservation into regional frameworks for urban growth, are both encouraged to be included.

In spite of the absence of UNESCO World Heritage site at Surat, the Archaeological Survey of India (ASI) has recognized six sites as "Monuments of national importance" in Surat. These are, the Dargah of Khawaja Dana Saheb Rouza, Old English Tombs, Tomb of Khawaja Safar Sulemani, Old Dutch and Armenian Tombs and Cemeteries, an ancient site comprising S. Plot No. 535 and the Fateh Burj (Udeaja et al. 2020)

From 2005 to 2014, the JnNURM (Jawaharlal Nehru National Urban Renewal Mission) program served as a stimulus for infrastructure development across the country with a special focus on the development of heritage regions. Jaipur, a historic city, is a prime illustration of the considerable efforts made possible by this plan. One of the first city level heritage plans in India to be integrated in the Master Plan of a city was the Jaipur Master Plan 2025, which was created to incorporate a Heritage Management Plan. This led to the creation of an action plan, a thorough heritage list, and an extensive analysis of the constructed historic resources.

The Masterplan for Delhi 2021 is another example of how heritage is included into masterplans. It identified heritage zones and archaeological parks and created Special Conservation Plans for listed buildings and precincts. Another technique to integrating planning, design, execution, and management is the creation of Heritage Management Plans (HMP) and City Heritage Cells (CHC). The World Bank's Cities Alliance, which was started at the government of India's request and runs from 2012 to 2018, recommended the creation of HMPs and CHCs as part of the revitalization of 40 historic cities. Jodhpur and Ahmedabad are used as examples. The HMPs and CHCs have carried out significant conservative interventions and activities.

Heritage walks and public education programs and events are some of the approaches used to raise awareness. In India, heritage walks have been a well-liked awareness campaign that have been successfully implemented in cities like Amritsar, Jaipur, Ahmadabad, Pondicherry, and Delhi. As a result, Ahmedabad citizens who reside in heritage homes made a concerted effort to clean both the exteriors of their residences and the surrounding historic district.

The study completed by the Urban Management Centre for the Surat Municipal Corporation served as the foundation for the survey. The Surat Central Zone and Rander Gamtal GIS maps were utilized to identify the sub-regions in the two historic areas with the highest number of historic structures, which will then be further researched on location according to the of state of conservation and real context situation. Close scrutiny of both of these areas helped the team better grasp some of the major obstacles to cultural preservation in Surat.

7.1 A Case Study of Pillayarpalayam Weavers' Cluster, Kanchipuram

Early Christians concentrated their cities at Kanchipuram, mostly due to its location along the banks of the Palar River, which provided the opportunity for trade. Weaving was regarded as a noble profession that produced silk clothing for the king and temples.

The Palar river, whose flow gave origin to a medieval water system in the city made up of canals, tanks, and surface water ponds, is the focal point of Kanchipuram's natural legacy.

The city, which has significant religious and cultural significance, has been selected by the Indian government to be among of the heritage cities under the HRIDAY Heritage City Development and Augmentation Yojana program.

The grid-like demarcated weavers zone primarily consists of residential land use. Figure 9 illustrates that traditional weavers' homes make up a sizable portion of the housing stock. The built environment is congested and devoid of space. This is primarily due to the merging of plots and the additions being constructed in the backyards of the homes. The design of the homes is reminiscent of the conventional Agraharam architecture. Temples are Agraharam's main attractions. The street is oriented from east to west. Common walls among neighbouring houses, a slanted roof with handmade baked clay tiles, wooden rafters, lime plastering, and an internal courtyard are some of the characteristics.

Silk saree weaving, a vital aspect of Kanchipuram's cultural history, is well-known. The Pillayarpalayam weavers are among the few indigenous weaving communities in Kanchipuram that continue to use traditional weaving methods and some of their original workstations. It was noted that a couple of the homes in the area

Fig. 9 Settlement analysis of Pillayarpalayam weaver's zone, *Source* (Vijayalaxmi and Arathy 2022)

did not experience significant structural changes, and the traditional workstations still house pit-looms and other traditional weaving tools. These homes have changed over time to meet modern requirements and family sizes. To understand how communities change, it is necessary to examine these ongoing changes and adaptations. The necessity of significant adjustments notwithstanding, the vernacular built heritage's authenticity to be maintained.

As a result, effective rules for the preservation and upkeep of the already-existing homes as well as the right building of new homes are necessary, and they must be implemented with the assistance of the local government, professionals, and the local community. Given that community involvement is essential to the preservation of vernacular heritage, appropriate information sharing between the stakeholders and the community should be assured. This would ensure sustainable growth for the craft as well as for the community by striking a balance between the preservation of the building form and community ambitions. Due to the vulnerability of the vernacular constructed heritage in this zone to change, it is necessary to propose policies and management plans through programs like INTACH and HRIDAY.

The forthcoming chapters in the book discuss conservation of private and public buildings. Conservation of private houses and palaces pose a separate challenge by itself. The conservation of the Chitnavis Wada in Nagpur is dealt in greater detail in the forthcoming chapter. Conservation of public buildings and temples have their own challenges. The strategies and Government intervention in implementing these strategies are discussed in this Chapter as well as in the other Chapters of the book with reference to the Chikkajala fort. Specific studies on conservation of Conservation of towns, and parts of towns are also important as some of the towns have historic significance or religious significance or defence significance. The strategies in general are highlighted in this Chapter. The further chapters of the book deals with specific case studies based on primary field studies and deals with the history, justification and conservation of historic towns such as Padmanabhapuram and Gwalior. Strategies for conservation include preservation, which has financial implications. Adaptive reuse as a strategy can be a win–win situation to the owner as well as the city. The use of the ARP method to assess the potential for adaptive reuse of the building is discussed in this chapter. Its application and use is demonstrated in further chapters with application in the Chettinadu house and the Victoria Building in Chennai. In order to ensure that the International Style of Architecture does not engulf the history and legacy of a place, public acceptance and desire for vernacular and historic strategies needs to be assessed. This will ensure that the flavour of history is retained in the contemporary buildings too.

8 Conclusion

India's historical past is eloquently represented in its ancient cities and public places, each telling a story of civilizations, cultures, and sociopolitical transformation. These historic cities and public spaces provide witness to India's complex fabric, embodying architectural marvels, urban planning genius, and the interplay between societal dynamics and physical surroundings. These urban places have not only served as centers of governance, trade, and cultural exchange but also as stores of collective memory and identity for the various people that have inhabited them.

The historical backdrop of Indian cities and public places reveals that they have played crucial roles in establishing the nation's narrative. From the well-planned towns of the Indus Valley Civilization, such as Mohenjo-Daro and Harappa, to the labyrinthine lanes of medieval cities like Delhi and Varanasi, India's urban landscape is a tapestry of indigenous cultures, colonial contacts, and global trade networks. Furthermore, these public areas have acted as stages for monumental events, social interactions, and cultural identity expressions. The unregulated dual processes of commercial conversion of residential properties and overpopulation with ensuing additional partition of domestic space were contributing to the widespread loss of traditional architecture, lifestyles, and urban shape. Courtyards were being covered over to make way for new godowns, markets, or apartments. According to one former high DDA official, all 5,000 katra courtyards under Delhi Development Authority (DDA) or Municipal Corporation control were subjected to similar changes. What was formerly known as an "Inverted Garden City" was on its way to becoming an industrial estate (Goodfriend 1982). Traffic congestion, infrastructural deterioration, unauthorised building, poor housing quality, and overall dreadful quality of life beset the walled city. The conversion of residential properties to non-residential purposes is a barrier to historical conservation, urban renewal, land assembly, valid property tenure, infrastructure upgradation, finance mobilisation, development rights, etc. The failure to implement plans resulted in it being designated as a slum under the Slum Area Act of 1956. Neither the legal framework nor the current organisational structure has been able to manage the unauthorised reconstruction of old Havelis/buildings, nor had they been able to undertake any substantial redevelopment. On the other hand, indiscriminate adoption of the various programmes has been counterproductive to the Special Area's economic and physical development. The case of increased urbanization, population expansion, and changing lifestyles hinders the preservation and restoration of historic cities and public spaces which has become a concern. However, increased awareness and desire to be rooted to ones heritage coupled with enormous Government support has shown encouraging results in the domain of Architectural Conservation in India. Non-destructive techniques such as of infrared thermography relies on the temperature of the materials being studied rather than any physical touch. Unreinforced masonry can be tested using ultrasonic velocity methods to assess a variety of attributes, including the uniformity of the brickwork, the locations of voids, cracks, and joints, and an estimation of the elastic properties of the masonry. These techniques are particularly helpful when examining historic buildings where the masonry must be conserved and destructive experiments are prohibited. Given that stone is used and that ultrasound testing is especially advised for this type of material due to the accurate and consistent results, the application of this process in India appears to be a viable option. The collection of methodically obtained digital photos of the site served as the starting point for the cloud-based digital photogrammetry process. Without the use of expensive or difficult equipment, digital two-dimensional pictures paired with readily available photogrammetric software make the process of documenting historic sites simple and straightforward. The automated modelling software (such as Autodesk ReMake) will process the photos after they are uploaded to the cloud-based server. As a result of the software's analysis of the two-dimensional photos, a three-dimensional polygonal textured mesh model is produced. This mesh model can be transferred to a computer and further processed there. Multiple mesh models may occasionally be independently developed and then combined. Using different modelling tools (like Maya or Rhino), the internal mesh model of a building can be added to the external mesh model of the building. Thus there are many non-destructive methods of assessing damage.

The Government is also taking keen interest in the preservation and conservation of historic structures and areas. Plans for Urban renewal, urban conservation, adaptive reuse etc. have been implemented. The huge number of structures can be overwhelming, but it is important to preserve and conserve the intangible and tangible cultural heritage to ensure that the legacy is passed on.

References

Bonazza A, Messina P, Sabbioni C, Grossi CM, Brimblecombe P (2009) Mapping the impact of climate change on surface recession of carbonate buildings in Europe. Sci Total Environ 407(6):2039–2050. https://doi.org/10.1016/j.scitotenv.2008.10.067

Brimblecombe P, Grossi CM (2008) Millennium-long recession of limestone facades in London. Environ Geol 56:463–471

Camuffo D (2019) Climate change, human factor, and risk assessment. In: Camuffo D (ed) Microclimate for cultural heritage. Elsevier, pp 303–340

Cassar M, Pender R (2005) The impact of climate change on cultural heritage: evidence and response. In: 14th Triennial meeting of the ICOM committee for conservation, 12–16 Sept 2005. The Hague, James & James, pp 610–616

Chandan S, Kumar A (2019) Review of urban conservation practices in historic cities. Int J Emerg Technol 10(1):74–84

Ciantelli C, Palazzi E, von Hardenberg J, Vaccaro C, Tittarelli F, Bonazza A (2018) How can climate change affect the UNESCO cultural heritage sites in Panama? Geosciences 8(8):296. https://doi.org/10.3390/geosciences8080296

Cutler NA, Viles HA, Ahmad S, McCabe S, Smith BJ (2013) Algal 'greening' and the conservation of stone heritage structures. Sci Total Environ 442:152–164

Dasra (2013) Crafting a livelihood, building sustainability for Indian artisans. www.edrfoundations.org

de la Fuente D, Vega JM, Viejo F, Díaz I, Morcillo M (2013) Mapping air pollution effects on atmospheric degradation of cultural heritage. J Cult Herit 14(2):138–145. https://doi.org/10.1016/j.culher.2012.05.002

Taher Tolou Del MS, Saleh Sedghpour B, Kamali Tabrizi S (2020) The semantic conservation of architectural heritage: the missing values. Herit Sci 8(70). https://doi.org/10.1186/s40494-020-00416-w

Goodfriend DE (1982) Shahjahanabad - Old Delhi: tradition and planned change. Ekistics 49(297):472–475. http://www.jstor.org/stable/43621804

Grøntoft T (2019) Observed recent change in climate and potential for decay of Norwegian Wood structures. Climate 7(2):33. https://www.mdpi.com/2225-1154/7/2/33

Guidelines for National Handicrafts Development Programme (NHDP) Office of the Development Commissioner (Handicrafts) (n.d.) www.handicrafts.nic.in. Accessed 26 Sept 2023

Haneef SJ, Johnson JB, Dickinson C, Thompson GE, Wood GC (1992) Effect of dry deposition of NOx and SO2 gaseous pollutants on the degradation of calcareous building stones. Atmos Environ A Gen Top 26(16):2963–2974. https://doi.org/10.1016/0960-1686(92)90288-v

Haugen A, Mattsson J (2011) Preparations for climate change's influences on cultural heritage. Int J Clim Change Strat Manage 3(4):386–401. https://doi.org/10.1108/17568691111175678

Howard AJ, Challis K, Holden J, Kincey M, Passmore DG (2008) The impact of climate change on archaeological resources in Britain: a catchment scale assessment. Clim Change 91(3–4):405–422. https://doi.org/10.1007/s10584-008-9426-9

http://heritageici.intach.org/wp-content/uploads/2020/03/2-Conservation-of-Flora-Fountain-Mumbai.pdf. Accessed 26 Sept 2023

http://heritageici.intach.org/wp-content/uploads/2020/03/9-Conservation-of-Lord-Surya-statue-at-Chhau-Kushinagar-Gorakhpur.pdf

http://inhcrf.org/ Accessed 26 Sept 2023

http://www.intach.org/. Accessed 26 Sept 2023

https://asi.nic.in/conservation-preservation/. Accessed 26 Sept 2023

https://www.tatatrusts.org/our-work/arts-and-culture/conservation. Accessed 26 Sept 2023

https://www.youtube.com/watch?v=7sBG8J2s-H8, The Stitches Speak

https://www.youtube.com/watch?v=RzTg7YXuy34, Sita Sings the Blues

ICOMOS Climate Change and Heritage Working Group (2019) The future of our pasts: Engaging cultural heritage in climate action. Outline of Climate Change and Cultural Heritage. Retrieved from https://indd.adobe.com/view/a9a551e3-3b23-4127-99fd-a7a80d91a29e

Kolay S (2016) Cultural heritage preservation of traditional Indian art through virtual new-media. Proc Soc Behav Sci 225:309–320. https://doi.org/10.1016/j.sbspro.2016.06.030

Lefèvre R-A (2014) The impact of climate change on slow degradation of monuments in contrast to extreme events. In: Schorlemer S, Maus S (eds) Climate change as a threat to peace. Peter Lang AG, pp 83–100

Menon A (2014) Heritage conservation in India: challenges and new paradigms. In: IX International conference on structural analysis of historical constructions at Mexico City, Mexico

Nagar PK, Gargava P, Shukla VK, Sharma M, Pathak AK, Singh D (2021) Multi-pollutant air quality analyses and apportionment of sources in three particle size categories at Taj Mahal, Agra. Atmos Pollut Res 12(1):210–221

NPC–AMASR National Policy for Conservation of the Ancient Monuments, Archaeological Sites and Remains (2014) Archaeological Survey of India Ministry of Culture Government of India Janpath, New Delhi 110011, pp 1–24

Orr SA, Young M, Stelfox D, Curran J, Viles H (2018) Wind-driven rain and future risk to built heritage in the United Kingdom: novel metrics for characterising rain spells. Sci Total Environ 640–641:1098–1111. https://doi.org/10.1016/j.scitotenv.2018.05.354

Petzet M (2004) Principles of preservation: an introduction to the international charters for conservation and restoration 40 years after the Venice Charter. In: International charters for conservation and restoration. Monuments and Sites, vol 1. ICOMOS, München, pp 7–29. ISBN 3-87490-676-0

Prieto B, Vázquez-Nion D, Fuentes E, Durán-Román AG (2020) Response of subaerial biofilms growing on stone-built cultural heritage to changing water regime and CO_2 conditions. Int Biodeterior Biodegradation 148:104882. https://doi.org/10.1016/j.ibiod.2019.104882

Sabbioni C, Brimblecombe P, Bonazza A, Grossi CM, Harris I, Messina P (2006) Mapping climate change and cultural heritage. In: Drdácký M, Chapuis M (eds) Safeguarding cultural heritage. Understanding & Viability for the Enlarged Europe, 2006. Institute of Theoretical and Applied Mechanics of the Academy of Sciences of the Czech Republic, Prague, Czech Republic, pp 119–124

Taylor J, Cassar M (2008) Representation and intervention: the symbiotic relationship of conservation and value. In: Saunders D, Townsend JH, Woodcock S (eds) Conservation and access: contributions to the London congress. IIC (International Institute for Conservation of Historic and Artistic Works), London, UK, pp 7–11. ISBN 0954816927

Types of Indian Handicrafts (n.d.) https://authindia.com/types-of-indian-handicrafts/. Accessed 17 Aug 2023

Udeaja C, Trillo C, Awuah KGB, Makore BCN, Patel DA, Mansuri LE, Jha KN (2020) Urban heritage conservation and rapid urbanization: insights from Surat, India. Sustainability 12(6):2172. https://doi.org/10.3390/su12062172

Vijayalaxmi J, Arathy KCK (2022) Critical evaluation of socio-cultural and climatic aspects in a traditional community: a case study of Pillayarpalayam weavers' cluster, Kanchipuram. Built Heritage 6(1). https://doi.org/10.1186/s43238-022-00050-4

Yaldiz E (2010) Climate effects on monumental buildings. Paper presented at the Balwois, Ohrid, Republic of Macedonia

Heritage Significance of the Historic Town of Padmanabhapuram, India, and Its Tourism Potential

Vidhya Joseph

Abstract *Padmanabhapuram,* originally known as *Kalkulam,* is an important historic town in the southern state of Tamil Nadu in India. Currently, the heritage significance of the town is primarily due to its palace complex dating back to the sixteenth century. However, this town was a part of the ancient Travancore kingdom ruled predominantly by the *Pandya dynasty* and *Ay dynasty* since the *Sangam period* (third century BC) and rose to prominence after it became the capital city of the *Venad kingdom* in the year 1553 CE. The town evolved as a capital over the period 1553 to 1795 and as a result, has a significant heritage of historic, natural, and cultural value. Today, the impact of urbanization and changing lifestyles is seen to have detrimental effects on the resources of heritage value at Padmanabhapuram. As suggested by SD11, protecting cultural heritage can foster policy coherence, driven by local and regional governments, involving increased citizen, and civil society participation. Therefore, tourism can be considered as one of the ways to raise awareness about the town to preserve its heritage. Primary field studies through listing, documentation, value assessment, and questionnaires are carried out to establish the heritage significance and assess the tourism potential. Sustainable tourism is deliberated to be the most appropriate model to raise awareness and preserve the rich heritage of Padmanabhapuram.

Keywords Historic town · Sustainable tourism · Padmanabhapuram · Heritage values

Glossary

Kottai	Fort
Marundhu	Medicine
Aaru	River
Kottaram	Palace

V. Joseph (✉)
Benny Kuriakose and Associates, Chennai, India
e-mail: vidhyajoean@gmail.com

Vijayalaxmi J. (ed.), *Conservation of Built Heritage in India*, Advances in 21st Century Human Settlements, https://doi.org/10.1007/978-981-99-8335-3_2

Chaava Kottai	Cremation fort
Kovil	Temple
Kulam	Tank/pond
Mandapam	Pavilion
Ottam	Procession (Run)
Gopuram	Entrance gateway
Naatu Vaidhyam	Native medicine
Balikal	Offering stone
Villaku	Lamp
Kodimaram	Flagstaff
Namashkara mandapam	Greeting pavilion
Vaathil	Door
Padipura	Gateway
Ottupura	Dining hall
Thai Kottaram	Queen's palace

1 Introduction

Historic urban areas, including cities, towns, and historic centres, serve as historic documents and embody traditional urban cultures' values through their natural and manmade environments (ICOMOS 1987). Indian historic cities are complex cultural resources, shaped by customs, lifestyles, materials, and socio-economic structures. They exhibit diverse morphological characteristics, functions, and hierarchies, and are organized through open, semi-open, and enclosed spaces, reflecting the heritage of different communities (Vijayalaxmi and Kalam 2022).

Kanyakumari district, known for its stunning scenery and favourable climate, is a significant tourist destination with numerous religious, social, cultural, historical, and environmental sites, but promoting tourism faces challenges due to a lack of information about its heritage sites (Gnanadhas and Usha 2014). The research categorizes tourist attractions into natural and cultural, but does not list all sites. It mentions Padmanabhapuram Palace and Vattakottai Fort but doesn't explain their connection to the heritage structures of the historic town. The town's numerous small forts suggest insecurity during that time (Nagam Aiya 1906). The Travancore State Manual provides an account of the fortification of the region; however, the defence system is not elaborated.

Kanyakumari abounds ancient monuments like forts and palaces. Important forts like Udayagiri and Vattakottai are very popular. Apart from these, there are many minor forts in the Kanyakumari district (Praveen 2018). The research provides a brief account of all the forts. However, an in-depth study in terms of its architecture, components, function, and inter-relationship between the forts is absent.

Oral traditions mention a tunnel connecting Charode Palace with the Padmanabhapuram complex. A tangible proof of this, perhaps, exists in Padmanabhapuram, where the door to a subterranean passage is still visible in the Thai Kottaram (Rajeev

2016). The article mentions tunnels in Kalkulam, a region with evergreen forests, but doesn't establish if they were part of a larger defence system. People use biodiversity for medicine through Ayurveda or Naatu Vaidyam (Silambarasan 2012). The paper brings out the medicinal value of the plants in the Kalkulam region of which Padmanabhapuram is a part, but does not consider it as a heritage asset of the natural value of Padmanabhapuram.

The ancient Tamils developed effective irrigation systems, with rivers serving as the backbone of irrigation in the Kanyakumari district, influenced by the Ay kings between the fourth century BC and ninth century AD (Santhosha Kumarai 2018). The research provides a historical overview of the Kodayar irrigation system in Kanyakumari district, but lacks an interpretation of its evolution and transformation into a modern system with numerous reservoirs and canals (Shoba 2018). The research paper examines water resources in Kanyakumari, Tamil Nadu, focusing on surface water quality for agricultural use in Padmanabhapuram Fort. It highlights pollution in ponds, which have continuous inflow from various sources, leading to a decline in water usage due to increasing contamination but neglects the historic water management system that existed to maintain the water resources (Ramesh et al. 2020).

Previous studies have not adequately explored the significance of Padmanabhapuram as a historic town, its function, and its morphology. The paper aims to highlight the importance of preserving this historic town while preserving its cultural and natural resources, and reinterpret its values while also analysing its tourist potential for sustainable preservation.

2 Study Area

2.1 Location

The historic town of Padmanabhapuram is located in the Kanyakumari district, a part of the southern state of Tamil Nadu in India (Fig. 1). It lies at the 80′15″ north latitude and 77′20″ east latitude and is situated about 15 km west of Nagercoil and 50 km east of Trivandrum. It has an altitude of 15 m above mean sea level and is bounded by Velli hills in the north-east.

2.2 Historic Background

Originally known as Kalkulam, Padmanabhapuram was a part of the Travancore kingdom. The earliest records of Travancore date back to the Puranic times, most importantly in the Kirthi yuga. Since the *Sangam age,* this region was ruled predominantly by the ancient *Panda dynasty* and briefly by *Chola.* Subsequently, it was ruled by the *Ay dynasty* (which had its roots in the *Pandya dynasty*) in the *historic*

Fig. 1 Map showing the location of Padmanabhapuram. *Source* Author

era. Padmanabhapuram became the capital of the Venad kingdom in 1553 AD and was later renamed Travancore. Padmanabhapuram was not the first stronghold of the Venad rulers, the villages nearby such as *Valliyur, Veerakeralaeswaram, Thiruvithamcode, Charode, Puliyoorkurichi (Udayagiri), and Eraniel* still have remnants of settlements, palaces, old fortifications, and major temples that predate Padmanabhapuram.

Under the reign of the celebrated king, *"Marthanda Varma" (1729–1758)* acclaimed as the "creator of modern Travancore", the kingdom extended up to central Kerala, and most of the heritage structures as we see today were built during that period (Fig. 2). The town flourished in the seventeenth century under the able leadership of Marthanda Varma who kept the forces of the Dutch and Tipu Sultan in check. The defensive fortifications were strengthened and the irrigation system was conserved between 1740 and 1744 with the help of Dutch general De Lannoy who was captured in the battle of Colachel. The proliferation of architecture during his period can be attributed to the economic boost brought about by the empire's expansion.

2.3 Boundary and Area

The area enclosed by the quadrangular fortification is about 186 acres. As of today, the region encompasses fertile agricultural wetlands, ponds namely Perumal Kulam, tanks such as Ramaswamy Kovil Kulam, Neelakandaswamy Kovil Kulam, Mudakulam, Krishnankovil Kulam, Saraswathi Kovil Kulam, Thekke Kottaram Kolam, Homapura Kulam, Padmanabhapuram Kottaram Kolam; settlements with old and new dwellings, temples, squares, processional routes, and market street within the fortified area. The Pattani Kulam (pond) lies outside the fortified area but forms a part of the town. The Valliaaru flows adjacent to the eastern side of the fortification.

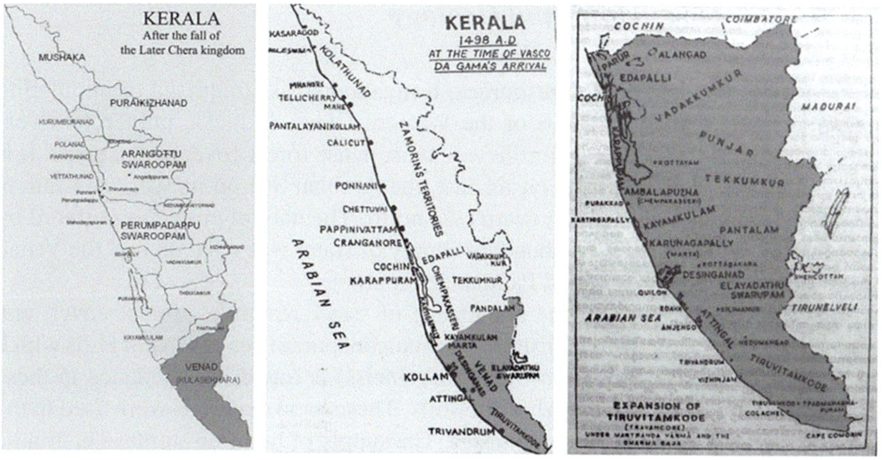

Venad Kingdom - 12'th CE Venad Kingdom - 15'th CE Venad rulers - Travancore Kingdom - 17'th CE

Fig. 2 Maps representing the expansion of the Kingdom under King Marthanda Varma. *Source* Archival maps of Kerala

The present municipal boundary however comprises an area of 6.47 square kilometres and includes part of Velli Hills, Marundhukottai Fort, Mayanakottai Fort, and numerous other water bodies (Fig. 3).

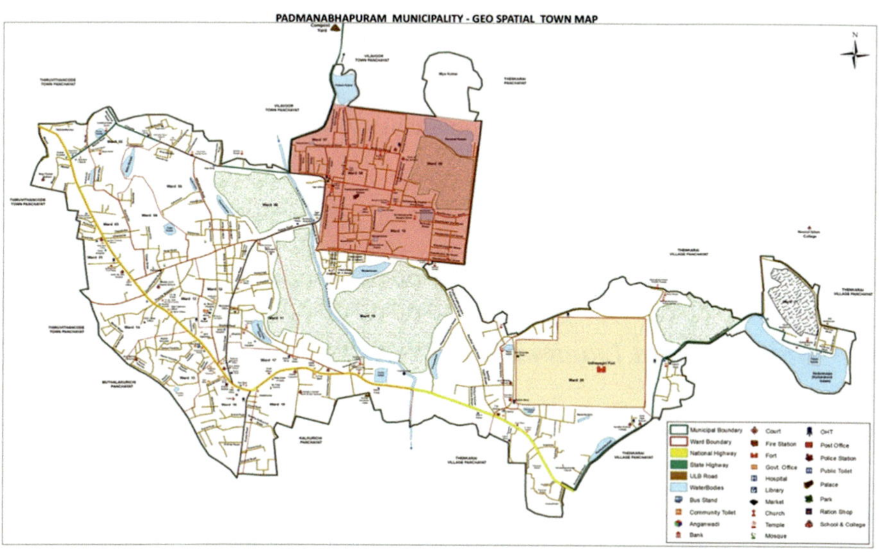

Fig. 3 Map highlighting fortified area within the municipal boundary of Padmanabhapuram *Source* Padmanabhapuram municipality

2.4 Geographic Setting and Geology

Blessed with abundant natural resources, Kanyakumari is comprised of undulating high mountains that form a part of the Western Ghats, hillocks, numerous rivers and rivulets, ponds, waterfalls, fertile wetlands, dense forest cover, and a coast. It is bounded by the Western Ghats on its east, the Arabian Sea on its west, the Indian Ocean on its south, and Kerala towards the north. The natural protection offered by the geographic setting and the abundant supply of water was what invited the Venad kings to establish their capital at Padmanabhapuram.

The maps (Fig. 4) reveal the abundance of water resources, forest cover, and mountains in the Kanyakumari district. Padmanabhapuram lies near Velli Hills which form a part of the Western Ghats. Granite (Gneiss) is found in abundance in these hills, while laterite content is high in the soils. These local materials were used in the construction of palaces, forts, and houses. The supply of building stones, i.e. granite was practically unlimited, but it was judiciously only used for the construction of temples and fort walls.

3 Methods and Methodology

3.1 Identification and Inventory

A detailed inventory of all the structures of heritage importance is carried out and is listed in the Architectural Heritage category of Table 1. The natural features unique to this place are listed under Natural Layers in Table 1. The aspects contributing to unique cultural landscapes are listed as Cultural Layer in Table 1. From Table 1 it can be seen that the entire area is rich in architectural heritage, natural and unique landscapes, and rich culture (Figs. 5 and 6).

4 Understanding: Components

4.1 Buildings and Landmarks

A1. Padmanabhapuram Fort

Padmanabhapuram town is surrounded by a quadrangular fortification that bounds an area of about 185 acres. The fort was constructed to protect the royal family and comprises the palace complex, temples, agricultural land, agrarian settlement, and water bodies within it. According to the Mutaliyar Manuscript of Alakia Pandiapuram, Padmanabhapuram Fort was planned in 1601 AD; before 1600 AD these forts were in mud. It took almost seven years to complete the fortification work.

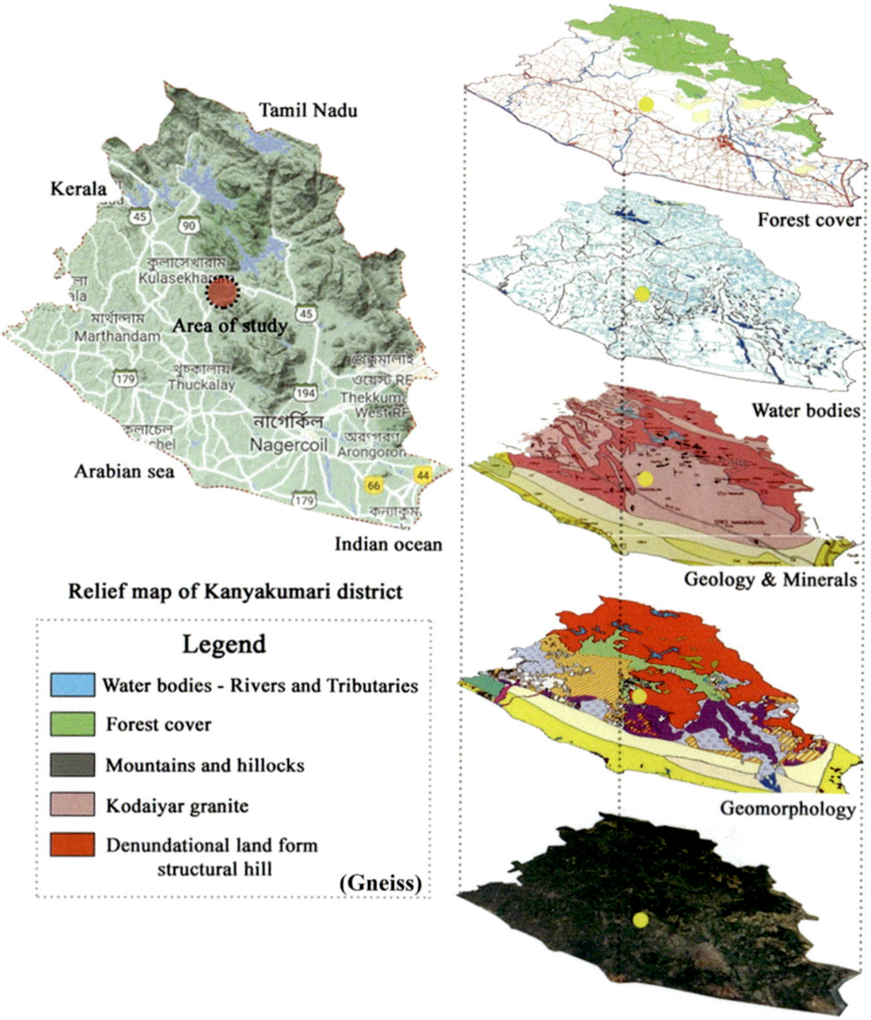

Fig. 4 Maps representing the geology and geography of Kanyakumari district *Source* Forest department Kanyakumari; Report on Kodaiyar river basin, District survey report on Kanyakumari; Google Earth

Sri Vira Ravi Varma was the ruler of Venadu (1592–1609) when the construction of mud fortification commenced. However, the fort was reconstructed into a granite fort in 1745 AD under the reign of Marthanda Varma. The fort was designed by Taikkattu Nambutiri, and the construction began in 1740 and went on till 1745 under the supervision of De Lannoy. The fortification included gates, bastion, and fort wall. The width of the fort wall is about 1 m and the height of the fort varies from 4.5 to 7.5 m, depending on the slope. The fort wall consists of two parts the rampart and battlement (equipped with merlons and crenelations). The rampart is made of

Table 1 Inventory of the Architectural, Natural, and Cultural Heritage of Padmanabhapuram town
Source Author

Architectural heritage layer				
S. No	Name	Location	Period of construction	Typology
1	Padmanabhapuram Palace	Padmanabhapuram	1550–1750	Residence
2	Padmanabhapuram Fort	Padmanabhapuram	1602–1745	Fort
3	Mayana Kottai—Cremation Fort	Puliyoorkurichi	1745	Fort
4	Marundhukottai—Fort of Medicine	Vilavoor	1744	Fort
5	Delannoys Tomb	Puliyoorkurichi	1777	Tomb
6	Udayagiri Kottai—Udayagiri Fort	Puliyoorkurichi	1597–1607; 1741–1744	Fort
7	Vattakottai Fort	Vattakottai	18th Century	Fort
8	South Travancore Lines	Aralvaimozhi Pass	1744	Fort
9	7th Shivalayam—Neelakandaswamy Temple	Padmanabhapuram	13th Ce–16th Ce	Temple
10	Ramaswamy Temple	Padmanabhapuram	1744	Temple
11	Devasagayam Pillai Church	Puliyoorkurichi		Church
12	Murugan Temple	Kumarakoil	6th Ce–8th Ce; 18th Ce	Temple
13	Eranial Palace	Eranial	15th Ce/16th Ce	Palace
14	Colachel Victorv Pillar	Colachel	1741	Port
15	Puthen Dam	Kulasehkaram	18th Ce	Dam
16	Padmanabhapuram Puthenar Canal	Padmanabhapuram	18th Ce	Canal
17	Pechiparai Reservoir	Pechiparai	1897–1906	Reservoir
Natural layer				
I	Velli Hills	Padmanabhapuram	–	Hills
2	Velli Aaru	Padmanabhapuram	–	River
3	Perumal Kulam	Padmanabhapuram	–	Pond
4	Pattani Kulam	Padmanabhapuram	–	Pond
5	Biodiversity Park—Medicinal	Pulivoorkurichi	–	Forest
6	Agricultural Fields	Padmanabhapuram	–	Farmland
7	Valli Chunai Waterfalls	Padmanabhapuram	–	Waterfall
Cultural layer				
	Name	Location		
1	Shivalaya Ottam—Shivarathri	Vilavancode (First Shivalayam)—Thirunattalam (L2th Shivalayam)		

(continued)

Table 1 (continued)

Architectural heritage layer				
S. No	Name	Location	Period of construction	Typology
2	Thirvathira	Padmanabhapuram Palace		
3	Navarathri Festival	1. Padmanabhapuram Palace—Padmanabhaswamy Temple 2. Kumarakoil Murugan Temple, Padmanabhapuram Saraswathi Temple, Suchindram Shiva Temple (Shiva)—Padmanabhaswamy Temple (Vishnu)		
4	Thirukalyanam Procession	Kumarakoil Murugan Temple		
5	Kavadi Kettu	Kumarakoil Murugan Temple		
6	Thaipoosam	Kumarakoil Murugan Temple		

Fig. 5 The tangible, intangible, and cultural heritage of Padmanabhapuram. *Source* Author, based on the municipality town map

granite blocks and the battlement is made of laterite blocks featuring a cyclopean masonry. There are 8 gates and they are made of granite blocks, and each of it varies in scale and ornamentation. One of the gates provides access to another hill fort known as Mayana Kottai. There are a total of 15 bastions; most of them are squares

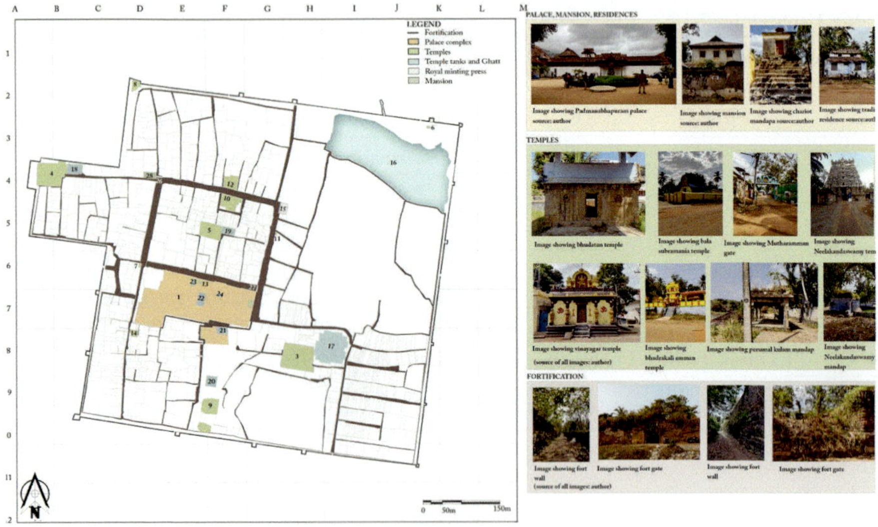

S.no	Name	Typology	Period of Construction
1	Padmananabhapuram Palace complex	Palace	1550 - 1750
2	Padmananabhapuram fort wall	Fort	1602 - 1745
3	Neelakandaswamy temple	Temple	13'th CE
4	Ramaswamy temple	Temple	1744
5	Perumal Temple	Church	1744
6	Utchi Pillayar temple	Temple	-
7	Vinayagar Temple	Temple	1744
8	Azhageshwari mutharamman temple	Temple	18'th CE
9	Mudakulam Azhageshwari mutharamman temple	Temple	18'th CE
10	Bala subramaniaswamy temple	Temple	1742
11	Theradi Bhudatan temple	Temple	18'th CE
12	Alampara bhagavathi Amman temple	Temple	18'th CE
13	Saraswathi temple	Temple	1745
14	Elupa Moodamman temple	Temple	-
15	Royal minting factory	Factory	1722
16	Perumal kulam Ghatt	Ghatt	-
17	Mahadevar koil kulam	Temple tank	13'th CE
18	Ramaswamy koil kulam	Temple tank	1744
19	Perumal koil kulam	Temple tank	1744
20	Mudakulam	Temple tank	-
21	Thekke kottaram kulam	Temple tank	1555
22	Neerali koil kulam	Temple tank	1750
23	Homarpura kulam	Temple tank	18'th CE
24	Saraswathi koil kulam	Temple tank	18'th CE
25	Community Wells	well	-
26	Mutharamman koil gate	Gate	-
27	Neelakandaswamy koil gate	Gate	-
28	Marudhanayagam pillai mansion	Residence	-
29	Heritage residences	Residence	-
30	Heritage shops	commercial	-

Fig. 6 Heritage mapping of built heritage in the fort area *Source* Author

and measure up to 9 × 9 m each. The bastions are made of granite and are designed with battlements (parapet wall). No traces of moat are found on 3 sides of the fort as they are either surrounded by agrarian land or defended by other forts (north side); however, the west-facing wall could have had a moat considering the level difference.

A2. **Padmanabhapuram Palace**

Located in the middle of the Padmanabhapuram Fort with the Velli Hills on the north-east, the palace was strategically placed considering the protection offered by the hills and the abundance of fertile farmland and water. The palace was constructed around

1601 CE by Iravi Varma Kulasekhara Perumal who ruled Venad between 1592 and 1609. The founder of modern Travancore, King Marthanda who ruled Travancore from 1729 to 1758, expanded the palace in around 1744 (Jacob and Ajay 2014). Padmanabhapuram Palace is considered as the largest wooden palace in all of India. Sprawling over an area of 6.5acres with 15 edifices, this elaborate complex boasts of intricate woodwork that throws light on the outstanding skills of the artisans and their traditional knowledge about the wooden construction. The Padmanabhapuram Palace complex consists structures such as Padipura (gateway), Poomukam, and Mantrasala: the King's Council Chamber; Thai Kottaram: Mother's Palace; Natak-sala: the Performance Hall; Upparika Maaliga: A four-storeyed king's mansion at the centre of the complex; Thekee Kottaram: the Southern Palace; Ottupura: Dining Hall, Plamootil Kottaram, Vepinmoodu Kottaram, Indravilasam, Puthen Kottaram, Homapura, Navarathiri mandapa, and Saraswathi Temple along with three temple tanks. The Mukhya vāthil, the principal gateway, is located on the western side of the fort wall and acts as the main entrance to the complex.

A3. **Shivalayam 7**

Neelakandeswarar Temple is located inside the Padmanabhapuram fortification. The temple has several layers as Travancore kings and Thirumalai Nayakar contributed to its construction and hence developed during the thirteenth–sixteenth centuries. It is a part of the famed Shivalaya Ottam conducted during Shivaratri and is the seventh temple out of a total of 12 temples. The temple has a mixture of both Tamil Nadu and Keralite styles and is one of a kind. It comprises a Gopuram, Mandapam, Kodimaram, and a huge water tank.

A4. **Ramaswamy Temple**

This temple was reconstructed in 1744 AD by Marthanda Varma, the Maharaja of Travancore. It comprises Srikovil, Namaskaramandapam, Nalambalam, Vilakku-matam, Balikkal mandapam, and a tiled superstructure built in typical Kerala style. The sculptural panels depicting the Ramayana story are part of the wall decoration of Vilakkumatam. The entire story is narrated in 46 wooden panels. They are placed in a circumambulatory direction as brackets on the full length of the four walls of Vilakkumatam (Nayar Preeta 2016).

A5. **Udayagiri Fort**

The Udayagiri fort was the principal arsenal of the Travancore Government, with a fine foundry, the only one it is believed to use in India for casting iron guns. The fort is at a distance of half a mile from Padmanabhapuram. Originally built from 1597 to 1607 by Vira Ravi Varma as a mud fort, it was rebuilt by Maharaja Marthanda Varma of Travancore in the eighteenth century (Fig. 7).

This fort was improved under De Lannoy's supervision during 1741–1744. Enclosing an area of about 90 acres (36 ha), including an isolated 260 feet (79 m) hillock, the fort contains the tomb of De Lannoy, who demolished the old fort and reconstructed it with granite walls and bastions. In the early days, the fort was of

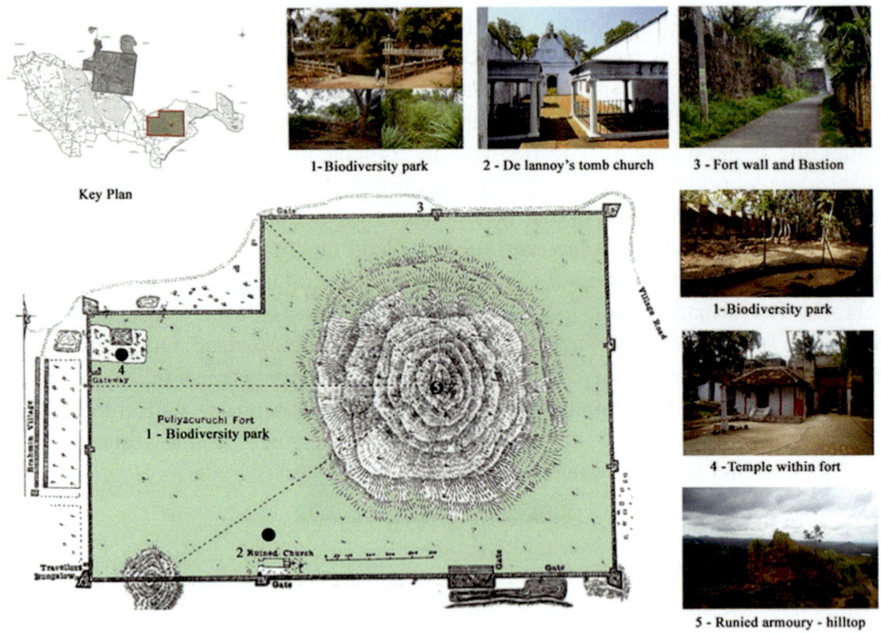

Fig. 7 Map and images showing De Lannoy's tomb, fort wall, and military structure *Source* Forts in Kanyakumari; Map based on an archival map in Travancore state manual

strategic importance. Prisoners captured in the campaign against Tipu Sultan were confined in the fort for some time. In later years, English East India Company troops were stationed at the fort until the middle of the nineteenth century. Presently, the fort is protected and has been turned into a bio-diversity park by the Tamil Nadu Forest Department, with sites of historical importance, such as De Lannoy's tomb, remaining as protected archaeological sites under the Archaeological Society of India. Recently, officials of the Department of Archaeology found a tunnel within the fort.

The various landmarks outside the fortification area are shown in Fig. 8.

A6. **Marundhukottai Fort of Medicine**

The fort of medicine was built during Marthanda Varma's reign under the supervision of De Lannoy in the 17th CE. The main purpose of the fort was to cure the wounded soldiers in times of war, and the presence of stone mandapa, hidden between boulders, testifies this. This hill-fort is is filled with medicinal plants and has an entrance on the eastern side with a granite frame. Today, the fort walls and the mandapam are in ruined condition. The wild growth of trees and their ramifying roots have done great damage to the fort walls and the mandapam. The fort wall and the gates are similar to Padmanabhapuram indicating that it was built during the same time.

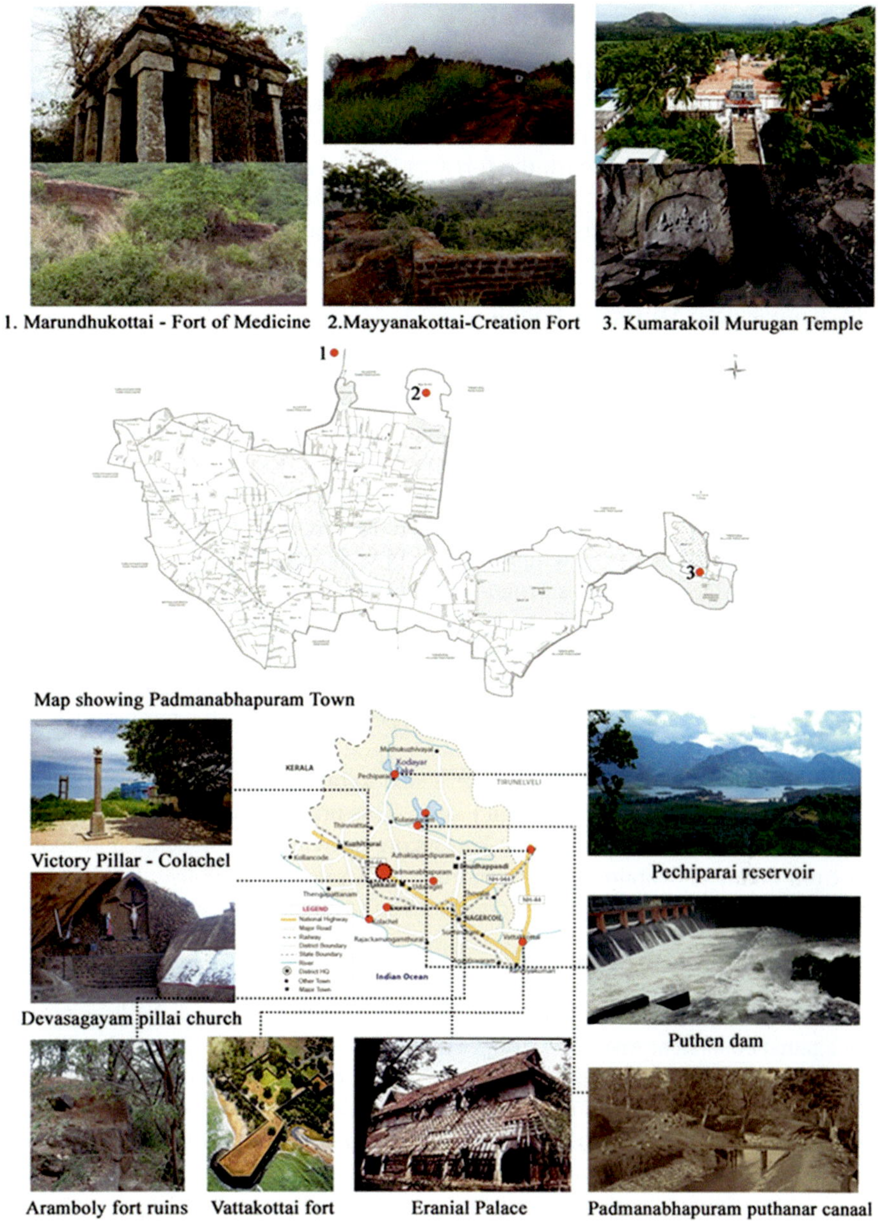

Fig. 8 Map and images showing the landmarks associated with Padmanabhapuram outside the fortification. *Source* Author

A7. **Mayana Kottai (Fort of Dead)**

Mayana Kottai is an ancient fort in Thuckalay, Tamil Nadu. The fort was built during the reign of Marthanda Varma by De Lannoy in 1700. It was built to bury the soldiers of the war. There were two rituals: either to burn the body or to leave it to eagles. To make the latter with the east, the fort was built at about 350 m above sea level. The fort was built of granite stones measuring 3′–5′ in length, 30–40 cm wide, and 25–30 cm thick. The ramparts of the fort have 44 gunpoints and 2 bastions. The total length of the fort was 165′ and breath was 95′. The entrance is 6′ tall.

A8. **Kumarakoil Murugan Temple**

Kumarakoil is situated at the Vellimalai hills in the Kalkulam Taluk. Velli Malai is believed to be the place where Lord Murugan married the tribal girl Valli. The time of construction of the temple is not clear but it also has a long history dating back to the sixth century. During the Tamil Sangam period, Kumarakoil was known as Thiruveragam as evidenced by Thirumurugattupadai, the Sangam literature written by poet Nakkeerar which testifies that the temple has evolved over the years. The inscriptions however date the temple to the eighteenth century.

It is an east-facing temple surrounded by paddy fields and coconut grooves and can be reached by climbing a fleet of 38 steps. Constructed in a blend of Dravidian and vernacular Kerala style, it comprises components such as Thirukulam, Rajagopuram, Mandapam, Ottupura, and Sannithanams (shrines) for several deities. There are separate enclosures for the sacred cows, bulls, peacocks, etc. in the temple enclosure (Praveen 2019).

During Navaratri Pooja, a procession is said to take place from here to the Devi Koil in the vicinity of the Trivandrum Palace. This practice has been in vogue since 1840 when the rulers of the erstwhile Travancore (province) changed their headquarters from Padmanabhapuram to Trivandrum. This practice demonstrates the strong connection between the erstwhile Travancore and Kumarakoil.

A9. **South Travancore Lines**

South Travancore lines also known as Aramboly lines was a wall built as protection against consistent invasions from Mysore during the rule of Tipu Sultan. It was constructed by Rama Varma under the supervision of his commander Eustachius De Lannoy during 1762–1775 CE. The lines by which the entrance into Travancore through the Aralvaimozhi pass was defended were about two miles in length stretching across the gap from one range of mountains to another. It consisted of small well-built bastions, joined at intervals by strong curtains; the whole fort was cannon-proof and protected by a thick hedge of thorny bushes. The fort commences from the Kanyakumari coast, passes through Murugan Kundram watch tower which is at an elevated level, then extends further to Kottiyadi and then to Kottaram. Another set of curtain walls extends perpendicular to the Aramboly line and joins the Vattakottai fort. The lines are almost washed out; only a few traces can be found at the coast, Aramboly forest and Murugan Kundram.

A10. **Vattakottai Fort**

The Vattakottai (circular fort) is a coastal defensive fortification built in the erstwhile Travancore Kingdom which now is a part of Kanyakumari. It was constructed in the eighteenth century by Punachal/Elakkara Valiyaveetil Marthandan Chempakaraman Pillai. Later, De Lannoy reconstructed Vattakottai, as part of the defence fortifications he undertook throughout Travancore. The fort is made of granite blocks and a part of the fort extends into the sea. The fort is now a protected site that can be visited by tourists.

A11. **Colachel Pillar**

The Colachel Pillar is located at Colachel town in Kanyakumari and was an important port town when the Venad Kingdom flourished. In 1741, a battle took place between the Dutch forces and the Travancore army. The Travancore army, though small in number, won the battle and as a memory erected a pillar at the battleground. As a result of the war, the naval commander De Lannoy was captured and later went on to become the military general of the Travancore kingdom. The pillar was built to commemorate the victory of Travancore over Dutch.

A12. **Eranial Palace**

Eranial was a seasonal capital of the Venad Kingdom until Padmanabhapuram became the capital. The origin of the Eranial Palace is still debated and some sources claim that it belonged to the fifteenth or sixteenth century. The palace extends up to three and a half acres and appears to be the predecessor of the much-acclaimed Padmanabhapuram Palace. Built in vernacular architecture, the palace is rich in terms of wooden details. The complex comprises the Padippura or the majestic entrance gateway leading to the main palace; the Kuthira Malika (Mansion of horses), the main area of the palace, which was a modest double-storied courtyard house; the Vasantha mandapam (Spring pavilion), a detached pavilion built on top of an elevated platform. It is replete with beautiful carvings. It is located on the western side of the palace complex. The palace compound also has a pond that was presumably used by royalty (Kolappan 2017). The palace is completely in ruins and only traces of it can be found. The palace is owned by the Government of Tamil Nadu, but poor maintenance has led to its present-day dilapidated condition.

A13. **Devasahayam Pillai Church**

Devasahayam Pillai was born into an affluent Hindu Nair family in Kanyakumari District. It was during Marthanda Varma's rule that Devasahayam Pillai and De Lannoy became well acquainted. De Lannoy's Christian faith interested Devasahayam and De Lannoy enlightened him on the faith, leading to Devasahayam Pillai's conversion in 1745. At this point, he took the name "Devasahayam". In 1752, the original order of the king and his dewan was to deport him from Travancore, into the Pandya country, at Aralvaimozhy. It was at Kattadimalai in Kanyakumari district that Devasahayam Pillai died on 14 January 1752. A church with historic remains was constructed at Puliyoorkurichi near Udayagiri Fort.

A14. **Pechiparai Reservoir**

Pechiparai Reservoir is the largest dam in the Kanyakumari district. It is located in the foothills of the Western Ghats in the Kanyakumari district of Tamil Nadu at Pechiparai. It was built during the period 1897–1906 by the European engineer Mr. Minchin by then Maharaja of Travancore. Due to the increase in the command area, the water available at Puthen Dam was found insufficient to meet the increased demand from the enlarging command area. Hence, the then Maharaja of Travancore decided to construct Pechiparai Reservoir across Kodayar.

A15. **Puthen Dam and Padmanabhapuram Puthenar Canal**

The Pandiyan anicut that existed since ancient times got silted up and became ineffective over the years. During the period of Venad King Marthanda Varma, a new anicut namely Puthen Anicut was constructed across the Paraliyar River about 400 m downstream of Pandiyan anicut. This anicut is a diversion weir, from which a canal called Padmanabhapuram Puthanar channel feeds the lands in Padmanabhapuram. Puthen Dam is the pivotal point regarding the irrigation system of the Kodayar basin.

B. **Squares and Street**

The fortified town comprises public and private squares, interactive spaces, and processional streets indicating the rich culture of the town. Several festivals associated with Hinduism happen in these squares throughout the year. The major streets include the north car street, west car street, and south car street where the procession takes place. The processions are taken on a wooden car/chariot which is manually pulled by devotees. The streets along which the procession moves are called the car street (Fig. 9).

The S-shaped road that leads to the palace comprises the major public square while the open ground (now encroached) in front of the processional road served ceremonial gathering purposes. The temple tanks that served ablution purpose is also used to house the festivities such as Shivaratri and Navaratri.

Rituals and Processional route: Some of the important processions and procession routes are the King's procession on the North, West, and East Car Street, the Perumal Temple procession, the Shilayam procession of 9 Shivalayas on Shivaratri, the Navaratri procession and Thirukalyanam procession which is conducted annually.

Fairs and festivals: Festivals such as Shivaratri, Thiruvadirai, Margazhi festival, Navaratri Utsavam, Vijayadasami Theppam, Panguni Uthiram Peruvizha, Thirukalyanam, and Thaipoosam are some of the important festivals. Fairs were arranged in the ceremonial gathering space indicated on the map which has been encroached upon today.

C. **Settlement**

The settlement pattern of Padmanabhapuram town is shown in Fig. 10.

Fig. 9 Map and image showing procession route, public squares, streets, and ceremonial gathering spaces in Padmanabhapuram town. *Source* Author; Map based on an archival map in Travancore state manual and Municipality town map

Analysing the figure ground of Padmanbhapuram, it can be surmised that it was not a greenfield development and had evolved organically over the years. The part indicated in yellow (Fig. 10) is the oldest part of the settlement that features a grid pattern as in any other agraharam. The humble agraharm village that developed into a capital city was planned as per Hindu planning principles of Artha shastra. After the shift of capital, the settlement lost its importance, and this led to the haphazard planning of settlement. The settlement today comprises traditional, modern, and semi-modern houses. There are very few houses with traditional elements such as sloped roofs, thick load-bearing walls, small windows and doors, and materials such as lime plaster, granite, mud, and wood. Commercial plots are found on the roads abutting the palace complex and appear to be recent as a result of tourism development.

The number of temples and shrines indicates that the settlement in the fortified area is primarily inhabited by Hindu communities. The paddy fields, coconut grooves, ponds, and river hint that the function of the settlement was agrarian. As the fort area is encroached currently, the plots are bifurcated; however, an old part of the settlement, as indicated in the map, shows traces of historic plot sizes with traditional houses that are only slightly modified.

Fig. 10 Map and image showing the settlement pattern of Padmanabhapuram town. *Source* Author; Map based on an archival map in Travancore state manual and Municipality town map

D. **Natural Components**

The site for the agrarian settlement was strategically chosen near the river Valli. It is a comparatively small river and runs 16 km long. It rises in the Vellimalai hills that bounds Padmanabhapuram. On both sides, paddy fields and coconut groves are found and enrich the rural economy of the area they pass. The river feeds the tanks that are used for irrigating the agricultural land in and around the Padmanabhapuram fortification (Fig. 11).

The ponds such as Kalkulam Shri Mahadevar Temple pond, having 4,780 m^2 area, Neerali pond with 798 m^2 area (private swimming pool for the royal family of kings), Mudakulam pond, with 602 m^2 area, Krishnankovil pond with an area of 100 m^2, Perumal pond with 28, 924 m^2 area, and Ramaswamikovil pond with an area equal to 620 m^2 are found inside the Padmanabhapuram fortification while the Pattani Kulam is found adjacent to the fortification. The tanks and pond are rain-fed, and the catchment areas align with the natural contours.

5 Understanding: Systems

The defence system and water systems are assessed.

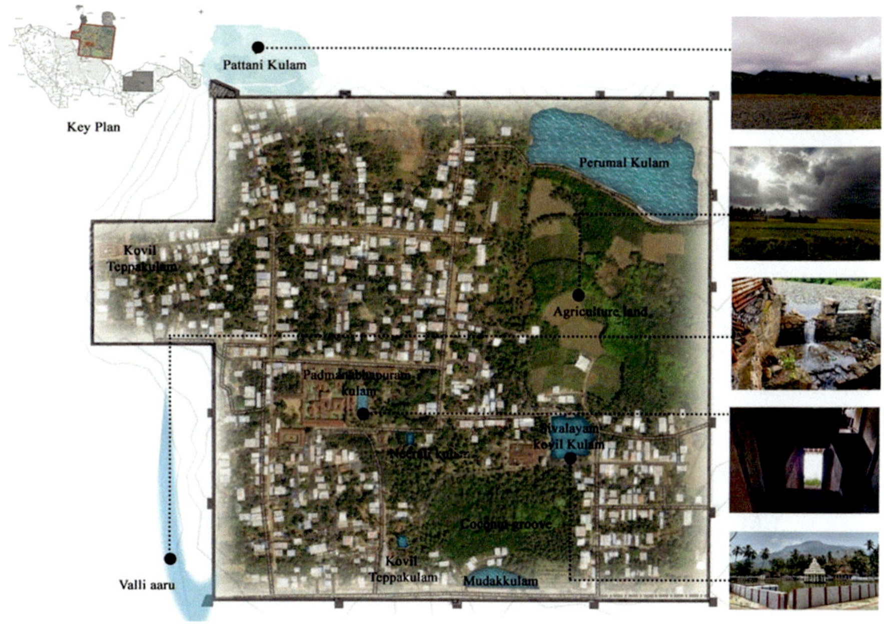

Fig. 11 Map and image showing natural resources in Padmanabhapuram town. *Source* Author; Map based on an archival map in Travancore state manual and Municipality town map

5.1 Defence System

The Travancore Kingdom extended between Kanyakumari and central Kerala and was bounded by the Indian Ocean in the south, the Zamorin Kingdom in the north (North Kerala), the Arabian Sea in the west, and the Western Ghats in the east. Although the Western Ghats are a series of high continuous hills, there exists a pass at Aralvaimozhi through which the feudal lords' neighbourhood Kingdoms invade the Travancore Kingdom.

The defence system of Travancore comprises two tiers of fortification (Fig. 12). Note: The lines indicated in the graphic only show the connections and do not show the route of tunnels. The first tier comprises the south Travancore lines that defend the empire, while the Padmanabhapuram Fort, Udayagiri Fort, Marundhukottai (Fort of Medicine), and Mayana Kottai (Cremation fort) form the second tier of fortification and defend the royal palace. The hill at Udayagiri fort was used as a watch tower to monitor the other forts, the palace complex, and the Indian Ocean through which battleships arrived. Besides the fort, a network of subterrain tunnels connecting Padmanabhapuram Palace, Udayagiri Fort, Charode Palace, and Eranial also existed

Fig. 12 Map and image showing the defence system of Padmanabhapuram. *Source* Author, based on GIS map prepared by Raj Bhagatt

for defensive purposes. The tunnels are not documented; only archival sources hint at the existence of these tunnels.

5.2 *Water System*

The main source of water is the rivers that are distributed through the natural and manmade canal which finally drains into the Indian Ocean and the Arabian Sea. A very good water management system existed since ancient times. Upper Kodayar is the main catchment that forms the primary source. Natural waterfalls fill the successive catchments (Fig. 13).

Dams like Puthen Dam and anicuts like Pandiyan Anicut divert the water, while reservoir like Pechiparai stores the water. Rivers such as Pazhayar, Kuzhithuraiyar, Chittar, and Valliyar are natural canals and artificial canals like Padmanabhapuram Puthenar flow through the kingdom and irrigate the agricultural fields. Lakes and ponds fed by these rivers are also used for agricultural purposes, while tanks which are rain-fed serve domestic purposes. The natural contour is studied and appropriate

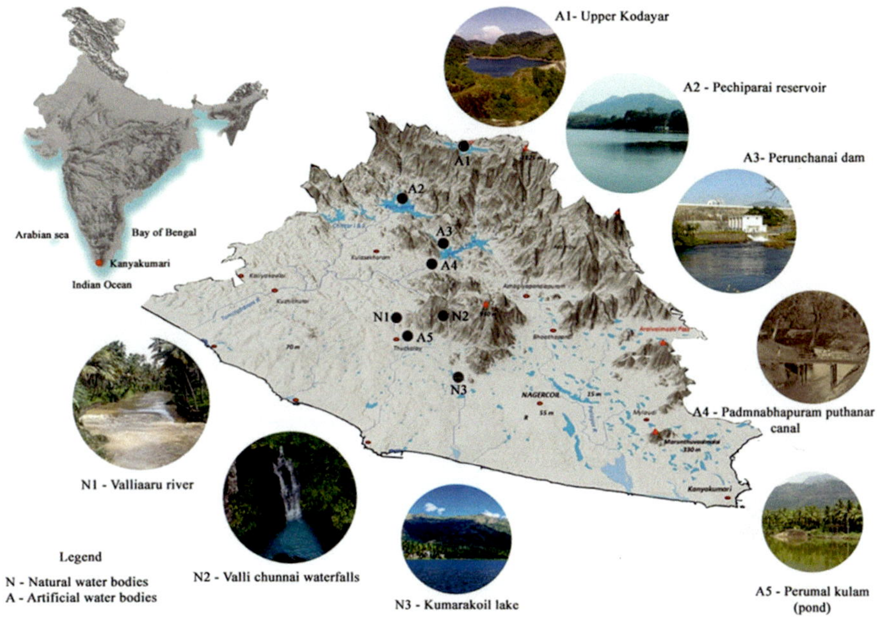

Fig. 13 Map and images showing the water system of Padmanabhapuram. *Source* Author, based on GIS map prepared by Raj Bhagatt

locations for these water sources were selected, hence the water flows through the natural slope and the intermediate recessions form the ponds and lakes.

6 Results and Discussion

6.1 Perception of Padmanabhapuram as a Tourist Potential

A questionnaire survey was conducted amongst 302 tourists who were largely people who visited the Padmanabhapuram Palace only. The survey was conducted to assess if the tourists were aware of the other touristic opportunities at Padmanabhapuram, the state of maintenance, and the reasons for neglecting the potential heritage areas. The results show that

- About 81% of tourists are from the states of Tamil Nadu and Kerala (adjoining state). So, it is mostly the locals who visit this place in majority.
- Adults form 66% of the visitors and 91% of the visitors are Heritage Enthusiasts.

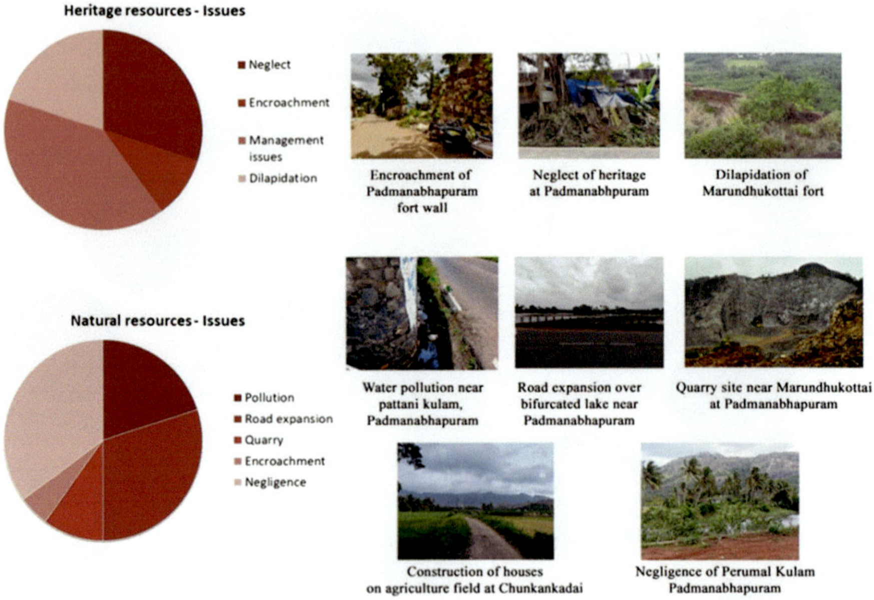

Fig. 14 Pie chart and images representing the threats issues of Padmanabhapuram *Source* Author

- 63% of the tourists felt that the buildings were in a state of abandonment and were poorly maintained.
- Apathy on the part of authorities is seen as the main reason for the neglect of important heritage buildings by 52% of respondents, while 28% felt that lack of awareness of the importance of heritage was the reason for neglecting.
- 42% of the respondents felt that inappropriate management of the place resulted in its poor condition (Fig. 14).
- Development in the form of road expansion and negligence together were perceived as a cause of neglect and loss of heritage by 68% of visitors.

6.2 *Value Assessment and Tourism Potential*

See Table 2.

Table 2 Table showing the Values and Tourism potential of Padmanabhapuram. *Source* Author

Inventory—value assessment and tourism potential							
Manmade layer							
S. No.	Name	Architectural style	Material	Integrity	Present condition	Value assessment	Tourism potential
1	Padmanabhapuram Palace	Vernacular	Wood, granite stone, brick, lime, terracotta	Structurally intact	Well-maintained	Historic, associative, aesthetic, cultural	Heritage tourism
2	Padmanabhapuram Fort	Vernacular—Colonial	Mud, granite stone, brick	Dilapidated	Encroached, Neglected	Historic, aesthetic	Heritage tourism
3	Mayana Kottai—Cremation Fort	Colonial	Mud, granite stone, laterite stone, brick	Dilapidated	Abandoned, Neglected	Historic, aesthetic	Heritage tourism
4	Marundhukottai—Fort of Medicine	Colonial	Mud, granite stone, brick	Dilapidated	Abandoned, Neglected	Historic, aesthetic	Heritage tourism
5	Delannoy's tomb	Colonial	Stone, brick, lime	Structurally intact	Maintained	Historic, associative, aesthetic	Heritage tourism
6	Udayagiri Kottai—Udayagiri Fort	Vernacular—Colonial	Mud, granite stone, brick	Structurally Intact	Maintained	Historic, natural, aesthetic	Heritage tourism
7	Vattakottai Fort	Colonial	Mud, granite stone, brick	Structurally intact	Well-maintained	Historic, aesthetic	Heritage tourism
8	South Travancore Lines	Colonial	Mud, granite stone, laterite stone, brick	In ruins	Abandoned, Neglected	Historic, aesthetic	Heritage tourism
9	7th Shivalayam—Neelakandaswamy temple	Dravidian—Vernacular	Granite stone, brick, lime, terracotta	Structurally intact	Living Temple	Historic, associative, aesthetic, cultural	Heritage tourism

(continued)

Table 2 (continued)

Inventory—value assessment and tourism potential							
Manmade layer							
S. No.	Name	Architectural style	Material	Integrity	Present condition	Value assessment	Tourism potential
10	Ramaswamy Temple	Vernacular	Granite stone, wood, brick, lime, terracotta	Structurally intact	Living Temple	Associative, cultural	Heritage tourism
11	Devasagayam Pillai Church	Contemporary	Stone, brick, concrete	Structurally intact	Living Church	Historic, associative, aesthetic, cultural	Heritage tourism
12	Murugan Temple	Dravidian	Stone, brick, lime	Structurally intact	Living Temple	Historic, aesthetic, associative	Heritage tourism
13	Eranial Palace	Vernacular	Wood, granite stone, brick, lime, terracotta	Dilapidated	Abandoned, Neglected	Historic, associative	Heritage tourism
14	Colachel Victory Pillar	Contemporary	Wood, granite stone, brick, lime, terracotta	Structurally intact	Maintained	Historic, associative	Heritage tourism
15	Puthen Dam	Colonial-Contemporary	Stone, lime	Structurally intact	Well-maintained	Historic, associative	Heritage tourism
16	Padmanabhapuram Puthenar Canal	Colonial	Stone, brick, concrete	Dilapidated	Well-maintained	Historic, associative	Heritage tourism
17	Pechiparai Reservoir	Contemporary	Stone, brick, concrete	Structurally intact	Well-maintained	Historic, associative	Heritage tourism
Natural layer							
1	Velli Hills	–	–	–	Encroached—Quarry sites	Historic, Natural, associative	Eco-tourism
2	Velli Aaru	–	–	–	Polluted	Historic, Natural	Eco-tourism
3	Perumal Kulam	–	–	–	Polluted—Shrunken	Historic, Natural	Eco-tourism

(continued)

Table 2 (continued)

S. No.	Name	Architectural style	Material	Integrity	Present condition	Value assessment	Tourism potential
Inventory—value assessment and tourism potential							
Manmade layer							
4	Pattani Kulam	–	–	–	Polluted—Shrunken	Historic, Natural	Eco-tourism
5	Biodiversity Park—Medicinal plants	–	–	–	Maintained	Historic, Natural	Eco-tourism
6	Agricultural fields	–	–	–	Encroached	Historic, Natural, Economic	Eco-tourism
7	Valli Chunai Waterfalls	–	–	–	Polluted	Historic, Natural	Eco-tourism
Cultural layer							
1	Shivalaya Ottam—Shivaratri	–	–	–	Living culture	Historic, associative, cultural	Community-based tourism
2	Thirvathira	–	–	–	Living culture	Historic, associative, cultural	Community-based tourism
3	Navaratri festival	–	–	–	Living culture	Historic, associative, cultural	Community-based tourism
4	Thirukalyanam procession	–	–	–	Living culture	Historic, associative, cultural	Community-based tourism
5	Kavadi Kettu	–	–	–	Living culture	Historic, associative, cultural	Community-based tourism
6	Thaipoosam	–	–	–	Living culture	Historic, associative, cultural	Community-based tourism

7 Conclusion

From the data collected and analysis made, it can be inferred that Padmanabhapuram is a significant historic town with historic, natural, socio-cultural, and associative values. SDG 11.4 suggests to achieve these goals, awareness-raising measures on diversity and inclusion, the promotion of intercultural and intergenerational dialogue, and the integration of cultural aspects in educational and lifelong learning strategies, which are fundamental. Different tourism models are required to be developed for natural, historic, and cultural heritage. Therefore, an integrated sustainable tourism model that benefits the economy, preserves the natural assets, and raises awareness about the heritage while preserving the integrity and authenticity of the historic town is found to be appropriate. Based on former research, an action plan is prepared to preserve Padmanabhapuram (Fig. 15).

Action Plan

It comprises six main steps:

1. The municipality of Padmanabhapuram to endorse the principles of sustainable tourism.
2. Establish a stakeholder group that includes the local communities of Padmanabhapuram.
3. Prepare a comprehensive integrated conservation management plan (including tourism, land-use planning, economy, etc.).

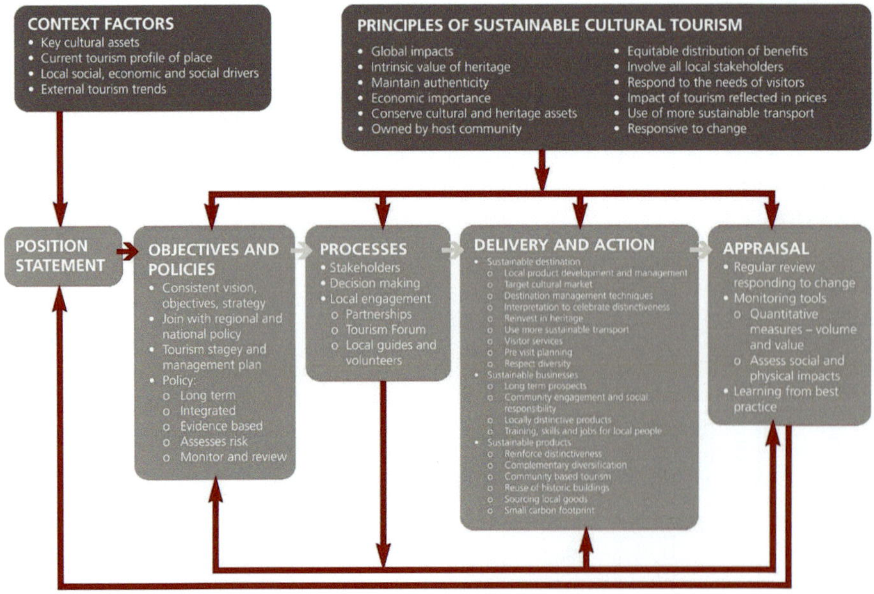

Fig. 15 Principles of sustainable tourism. *Source* Based on Brian Smith and Matthias Ripp (2020)

4. Open and honest assessment of the current position of the heritage sites and compare it to Guidelines for Sustainable Cultural Tourism in Historic Towns and Cities.
5. Agree on actions, resources, and timetables to reflect the local circumstances and needs.
6. Implementation of the action plan.

References

Smith B, Ripp M (2020) Guidelines for sustainable cultural tourism. Organization of World Heritage Cities. https://www.historic-towns.org/wp-content/uploads/2020/07/Guidelines-for-Sustainable-Cultural-Tourism-2020-1.pdf

Gnanadhas E, Usha A (2014) A study on sustainable tourism development in Kanyakumari district. Asia Pac J Mark Manag Rev 3(5):1–6

ICOMOS (1987) Charter for conservation of historic towns and urban areas. As read from https://www.icomos.org/images/DOCUMENTS/Charters/towns_e.pdf on 23'rd may 2023

Jacob TG, Ajay K (2014) Unfolding creative order: a study of Padmanabhapuram palace, as read from https://repository.cept.ac.in/handle/20.500.12725/6687?mode=full on 23'rd may 2023

Kolappan B (2017) Eraniel palace beyond restoration. The Hindu

Nagam Aiya V (1906) The Travancore state manual. World Public Libr Assoc 1:200–207; 237–307

Nayar P (2016) Animation technique and depiction of death in sculptural art: ramayana story in Shri Ramaswamy temple in erstwhile Travancore, Heritage. J Multidiscip Stud Archaeol

Praveen OK (2018) Important forts of Kanyakumari, Laxmi book publication—Impact Factor: 1.9152(UIF), 4(5):1–5

Praveen OK (2019) Kumarakoil Murugan temple—a study. Online J Multidiscip Subj 13(2):140–141

Rajeev SS (2016) Depths of mystery, The Hindu, as read from http://sharatsunderrajeev.blogspot.com/2016/ on 23'rd may 2023

Ramesh BK, Velayutham Pillai M, Vanitha S, Diagu J (2020) Analysis of surface water quality for irrigation in Padmanabhapuram fort. IOP Conf Ser Mater Sci Eng 872:1–6. https://doi.org/10.1088/1757-899X/872/1/012191

Silambarasan G (2012) Herbal Folk Medicine used Against Rheumatic Problem and Jaundice in Kalkulam Taluk. Inven Rapid: Ethnopharmacol 2012(4):1–3

Santhosha Kumarai H (2018). Early irrigation system in Kanyakumari district, International Journal of Creative Research Thoughts, Vol 6, Issue 1, pp 876 - 878

Shoba C (2018) Water resource system in Kanyakumari district. J Emerg Technol Innov Res 5(6):567–572

Vijayalaxmi J, Arathy Kalam KC (2022) Critical evaluation of socio-cultural and climatic aspects in traditional weavers' community—a case of Pillayarpalayam weavers' cluster, Kanchipuram. Built Herit 6:1. ISSN: 2662-6802

The Lost Heritage of Bengaluru City—The Case of Chikkajala Fort

Hrudya Seemala

Abstract The metropolitan city of Bengaluru has been undergoing rapid urbanization for the past two decades. The influx of Information Technology has taken over in the form of an IT hub. As a result, the essence of heritage has been taken aback. This study pertains to the heritage structure, Chikkajala Fort, located on the outskirts of Bengaluru Urban. The city has undergone the consequences of urbanization. Chikkajala Fort is a temple complex with a history dating from the Iron Age (1,000BCE), Hoyasalas (eleventh to fourteenth century), Vijayanagara (fourteenth to seventeenth century), Tipu Sultan and Hyder Ali (seventeenth to eighteenth century), to Chettiyars during the nineteenth century. Its significance is also weighed by the values it possesses such as historical, architectural, socio-economical, and cultural values. Considering its values, there is a need to strategize its conservation since it is in danger of succumbing to urbanization. This paper aims to establish the associative values and their significance to arrive at conservation strategies for its protection. The study is mainly conducted based on qualitative research techniques: documentation on the field, primary field survey, interviewing the locals to understand the socio-cultural structure and practices, etc., which helps in interpreting and analysing the area to arrive at a conservation strategy.

Keywords Urbanization · Heritage · Historic aspects · Values · Chikkajala fort · Conservation

1 Introduction

The built heritage is the fundamental expression of the culture of a community, and its relationship with its territory (Vijayalaxmi and Arathy 2022). Bengaluru, the capital city of Karnataka, is located in the southern part of India. In recent times, there has been a lot of growth and development in the city, resulting in a gap between the past and present. Bengaluru city has a history dating back to about 3,000 years ago. The

---

H. Seemala (✉)
School of Planning and Architecture, Vijayawada, India
e-mail: hrudya98@gmail.com

Vijayalaxmi J. (ed.), *Conservation of Built Heritage in India*, Advances in 21st Century Human Settlements, https://doi.org/10.1007/978-981-99-8335-3_3

Chikkajala Fort remains are one of the historic places that take back to the time of Megalithic tombs which are of the Iron Age. The Chikkajala Fort is located about 15 km away from Bengaluru, in the precinct of Chikkajala village. The fort is a temple complex that consists of a small shrine, a stepwell (*Kalyani*), and two pillared halls (*Dharmachattra*) which were initially constructed during the eleventh century by the Hoyasalas and then had later interventions in different significant periods. This type of fortified temple complex is unique on its own and talks about the historicity of the area, its architectural characteristics, and the community and culture associated with it. The features and the values it holds give a cue on how the area had been established and had been used over the period. As suggested by SDG 11, one way that protecting tangible and intangible cultural property can promote bottom-up policy coherence is through increasing participation from citizens and civil society, headed by local and regional governments. The fort area during its period under the Archaeological Survey of India (ASI) yet had been a bait to the National Highway Authority of India (NHAI). In the year 2012, a part of the fortification was destructed for road widening at NH-7, on the way to the airport as a result of urbanization. Presently, due to the disputes by owners, the fort complex is no longer under the ASI and has undergone a lot of other encroachments as well (Gogoi 2018). The study is mainly to establish and signify the historic, cultural, architectural, economic, and socio-economic values and analyse them to come up with a strategy to conserve them.

2 Study Area

2.1 Site and Its Surroundings

Bengaluru, also known as the Silicon Valley of India, is the capital city of Karnataka which is located in the southeast part of India (Fig. 1). It is located at 12°58′44″N 77°35′30″E and covers 741 km^2 (286 sq. mile). It is at a height of over 900 m (3,000 ft.) above sea level, on the Deccan plateau, thus creating pleasant and moderate climatic conditions. Through time, the city has undergone rapid growth and urbanization blurring its historicity.

One of the lost heritages of the city, as per the scope of the study, is the Chikkajala Fort. The fort ruins are located on National Highway-7 on the way to the airport, in the vicinity of Chikkajala village, which is 15 km away from Bengaluru, Karnataka. It is about 2 acres and is a small temple complex. It is surrounded by the National Highway in the west, shopping complex to its north, and east side consists of open fields and Akkayamma betta. This is the north part of Bengaluru, which is the sub-urban area undergoing rapid development (Fig. 2). The location is such that it has easy access visually and circulation-wise.

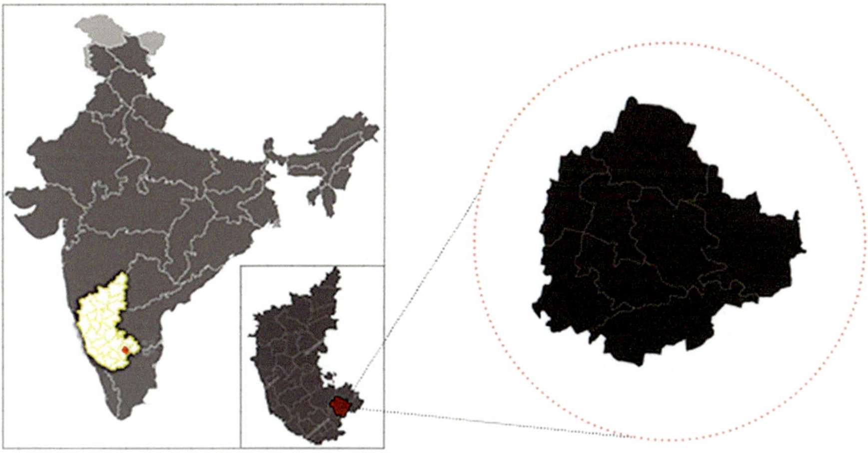

Fig. 1 Location of Bengaluru. *Source* Author

Fig. 2 Location of Chikkajala fort complex. *Source* Author

2.2 *History*

2.2.1 Bengaluru

The city dates back to prehistoric times. The most prominent rule of the area began from the Western-Ganga Dynasty 350-1000AD, since then ruled by other empires such as the Cholas (300BCE-1279), Chalukyas (543-753CE), Hoyasalas (c.1050-c.1355), Vijayanagara (1336–1646), and so on till the time of British colonial rule (1858–1947). The city was officially founded by Kempe Gowda in the year 1537, who was the chieftain (Nada Prabhu) of the Vijayanagara Empire (Abhishanker 1990). The Nada Prabhu developed the city over time by building forts, temples, Kalyanis, etc. There was a recent discovery of a burial ground dated back to 1,000 BCE on the outskirts of Bangalore at Chikkajala Fort premises by the ASI. An excerpt states "In the open fields on the eastern and southern sides of the village are located some cist burials. The cists are enclosed by irregular boulders some of which are missing. Some of the disturbed burials have yielded black-and-red ware, human bones, and a few iron implements." which was found as evidence of the existence of the prehistoric layer (Priya 2020).

2.2.2 Chikkajala Fort

The Chikkajala fort is about 2 acres consisting of a small temple shrine of lord Hanuman, a stepwell which is known as *Kalyani* in the region, and two huge pillared halls known as *Dharmachattra* (Fig. 3). Firstly, in terms of historic layers, as mentioned it dates back to the Iron Age which is around 1,000 BCE, the time which mentioned the existence of the megalithic tombs. Each component of the complex was built during different periods. It began with the Hoyasalas in the tenth century where the Palegars had built the fort wall and a temple which is now destructed due to wars and only a part of it is left. Then it was taken over by the Vijayanagara Empire in the fourteenth century.

By the eighteenth century, Hyder Ali had made significant additions to the fort wall. Over time the fort wall had been destructed and had to be rebuilt. It was during the Tipu Sultan period it had been rebuilt thus having a similarity with the Devanahalli Fort which is from the same period, located a few kilometers away. The presence of the fort in a temple complex can talk about the cultural and spiritual significance it holds. By the nineteenth century when Bengaluru city was in the Mysore province, the Chikkajala fort property was bought by Doddi Appan Chettiar in the year 1859 through auction. The Chettiyars built the Kalyani as a later intervention in front of the Hanuman temple shrine. The temple complex was taken care of by the Chettiyars till 1958, then they handed it over to the state government.

2.3 Socio-economic and Cultural Aspects

The temple consists of the Hanuman deity, which is in the form of a stone relief. It has historical evidence that dates back to the Hoyasala Empire (eleventh to fourteenth century). It was the reign where the historic sects of Shaivism and Vaishnavism both co-existed. The worshippers of Shiva were Shaivas or Lingayats, and the worshippers of Vishnu are called Vaishnavas. This temple was built by the King who was an ardent follower of Vishnu. It can be inferred that the community visiting the temple follows Vaishnavism Bhargava (2010). Vaishnavism is under the Hindu sect where the incarnation of Vishnu (Rama, Sita, and Hanuman) and Lord Vishnu himself was worshipped. People from Devanahalli, Chikkajala, Chikkabalapur, and Doddabalapur villages used to come for worshipping and to give offerings.

Fig. 3 Chikkajala fort complex. (i) (**a**), (**b**), (ii) (**c**), and (iii) (**d**). **a** Chikkajala Fort complex—(i) Stepwell (Kalyani) and Hanuman temple, (ii) Pillared hall, and (iii) Fortification. *Source* Author

The temple complex was initially constructed to have various functions, marriages, and also as a place of a halt during the nighttime. The economic aspect was derived from the functions and was used for the requirements. As of now, to keep the tradition and spiritual aspects alive, a priest comes from the nearby village, Devanahalli, for the rituals.

2.4 Issues

At present, the front part of the complex which is on National Highway-7 has been demolished by the National Highway Authority of India (NHAI) in the year 2012, to widen the road. But over a decade, there had been no action taken on restoring or repairing it.

Over time, it has just been neglected and a huge shopping complex has been constructed right next to it. Apart from these larger interventions that had encroached on the fort complex, there are other issues within it such as excessive growth of vegetation, dilapidated structures (pillared halls), and absence of water in the stepwell (Kalyani) (Fig. 4).

Fig. 4 Time series Google Earth images of Chikkajala Fort in the course widening of NH-7. *Source* Gupta et al. (2017)

3 Spatial Characteristics

The temple fort complex which is about 2 acres consists of 4 main components: (a) Stepwell (*Kalyani*), (b) Hanuman temple, (c) Pillared halls (*Dharmachattra*), and (d) Fort wall (Figs. 5 and 6). Over time, these spaces have gone through additions and alterations. It thus serves as an evidence of the historical layers, architectural styles, and the construction techniques.

3.1 (a) Stepwell (Kalyani)

Wells or reservoirs constructed as a part of a temple complex next to an Indian temple are known as "temple tanks" or "Stepwells." A stepwell is a deep masonry well with stairs leading down to the well's water level. It is known as *Kalyani* or *Pushkarini* in the south (Fig. 7). They are ancient Hindu stepped bathing wells, and when bathed in, some tanks are believed to treat various illnesses and maladies. A *Kalyani* is typically rectangular or square in shape.

The Chikkajala temple complex has a square-shaped Kalyani which is 95ft. × 95ft., and is almost 25–30 ft. deep which is stepping down to the water source. It consists of four entrances in four cardinal directions. It is made of Granite, which is a non-porous material and can hold water within for a long time. By analysis, it can be said that it is a stepwell that depends on rainfall for it to be filled and also by percolation of groundwater. In the present scenario, it is dry and filled with vegetation.

Fig. 5 Chikkajala fort complex (3D view). *Source* CMRUSOA, Bengaluru Batch-2019

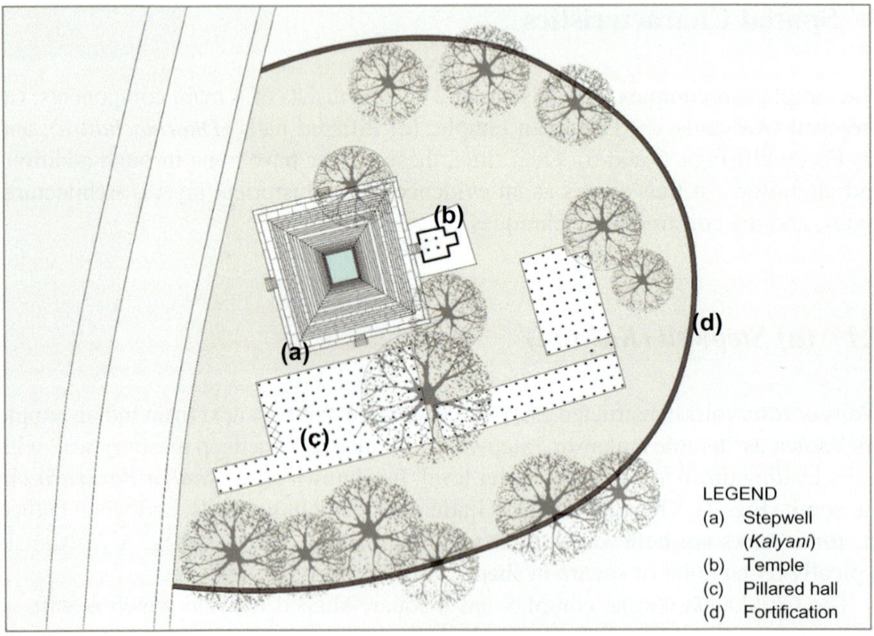

Fig. 6 Plan showing Chikkajala fort complex. *Source* Author

Fig. 7 Kalyani. *Source* Author

Fig. 8 Fish. *Source* Chikkajala Fort (2015)

The *Kalyani* can be dated back to the tenth century which was during the Hoyasalas. The Kalyani is a classic example of the Hoyasala architecture. It is a testament to the ancient Indians' skill in managing water because these wells not only held water but also allowed groundwater to percolate through them. The steps of the well have some relief carvings of fish, a bird carrying a twig, childbirth, etc. (Figs. 8, 9, 10 and 11).

3.2 (B) Hanuman Temple

Right in front of the Kalyani towards the east is the temple shrine which is devoted to Lord Hanuman. The deity which can be observed is by an old stone of Hanuman's relief sculptor (Fig. 12). The temple consists of a small *Mantapa* and a *Garbagriha* where the deity is kept. It is a small structure that is about 16ft. × 16ft. The *Shikhara* consists of intricately done work with 3 *kutas* and *shalas* on yellow sandstone dating back to the Hoyasala period, and later it had been altered by the Mysore kings by adding Mysore stucco work on it. It was during the rule of the Hoyasala Empire which was around the eleventh to fourteenth century that there was a vigorous temple-building activity which was due to social, cultural, and political events of the period. The stylistic transformation of the *Karnata dravida* temple-building tradition reflected religious trends popularized by the Vaishnava and Virashaiva philosophers. This temple was built during the reign of King Vishnuvardhana; he was very inclined

Fig. 9 Childbirth. *Source* Chikkajala Fort (2015)

Fig. 10 Bird carrying a twig. *Source* Nast, n.d.

to Vishnu and his followers were Vaishnavites. This can be seen even in the presence of Lord Hanuman the deity who is worshipped by the Vaishnava faith.

At present, the temple has a shikhara with intricate details and a shrine that dates back to the thirteenth century with the substructure having a very simple form with no intricate carvings.

Fig. 11 Elephant. *Source* Nast, n.d.

Fig. 12 Hanuman temple. *Source* Kirana Bhat (2020)

3.3 (C) Pillared Halls (Dharmachattra)

There are two pillared halls, towards the south and east of the *Kalyani* (Figs. 13 and 14). They are built out of stone in a trabeated manner consisting of long rows of pillars. It is believed to be 300 years old. It is known as the Dharmachattra in the local language. Usually, at any pilgrim area, a space for a halt is given. This is built as a night resting space for travellers and cattle from surrounding villages and towns.

Fig. 13 Pillared hall towards east. *Source* Author

Fig. 14 Pillared hall towards south. *Source* kiranabhat.com (2020) Author

The halls towards the south of the Kalyani are rectangular-shaped areas with about 178 pillars that are trabeated, which means there is a use of simple lintel and beam support with a roof. These are built using granite and sandstone which can be inferred by observation. The main structure which consists of the pillars are made of granite stone, and the superstructure which has a band-like appearance that is placed on the beams are made of sandstone. It also has intricate carvings on them (Fig. 15). From observations, it can be analysed that there was a room-like enclosure that was done using bricks and now a large part of it is dilapidated. A banyan tree is seen growing from the stone platform creating structural deformities.

The pillared hall towards the east of Kalyani is right behind the temple. It is a smaller structure and has similar features as the others described.

Fig. 15 Pillar. *Source* Kirana Bhat (2020)

3.4 (D) Fort Wall

A fort wall is made to protect the area enclosed within it. It is a type of defensive system adapted to prevent destruction. Most of the fort walls are made such that they can withstand attacks from rivals. The most peculiar part of the Chikkajala fort wall is that it's not made strong enough to withstand attacks (Fig. 16). It was initially made by the *Palegars* (local kings) during the rule of Hoyasalas. Over time, it was being attacked and there was some destruction. By the fourteenth century, it was in the hands of the Vijayanagara Empire. By the eighteenth century, it was during Hyder Ali's rule that the fort had been rebuilt and it has similar features as the Bangalore Fort which is located in the central part of the city away from the Chikkajala area, which was built and altered during the same period. They used stone and masonry for the fort. Initially, the fort was an enclosed area with a main entrance towards the east. It consisted of a gate (Fig. 17) made of sandstone with intricate carvings. It was a flat-roofed gate with three openings, the middle being the main entry and security enclosure on either side. The superstructure consists of relief structures of elephants, gargoyles, etc. as a grandeur entrance.

Fig. 16 Fort wall. *Source* Jagannath, U. (2011)

Fig. 17 Fort gate entry before demolishing. *Source* Vijayalaxmi and Arathy (2022)

4 Materials and Methods

The study aims to document the significance of Chikkajala Fort through field documentation, surveys, interviews, and data analysis. It benchmarks the site against similar heritage sites to identify best practices. Conservation strategies are developed based on research findings and policy assessment. The study also evaluates existing conservation policies to identify gaps and areas for improvement. Ethical considerations are addressed. The research aims to raise awareness about Chikkajala Fort's significance, involve the local community in its preservation, and propose strategies for its protection in the face of urbanization. Tourism revenue is suggested to support ongoing maintenance efforts. The research aims to bridge the gap between historical importance and contemporary challenges.

5 Changes Over Time

From the figure (Fig. 18), we can observe that the site has undergone a lot of changes over time. Initially having an intact enclosure (Fig. 18a) in the year 2010, by 2012 the front gate has been destructed by the NHAI (Fig. 18b). At present, in the year 2022 (Fig. 18c), there is a new shopping complex that has been constructed beside it. Apart from the growth around it, the site has also been through a lot of vandalism and had been excavated by goons for finding any artefacts.

On the positive side, as it has a living temple, it is used for worship, and throughout the day nearby workers and auto drivers take rest under a tree. Many people are also becoming aware of the place and are visiting it. The values associated with it make the people from the city come forward to create awareness.

5.1 Present Condition

At present, with the ongoing urbanization, the Chikkajala fort complex is facing various issues. With growing developments and changes, once a prehistoric site under

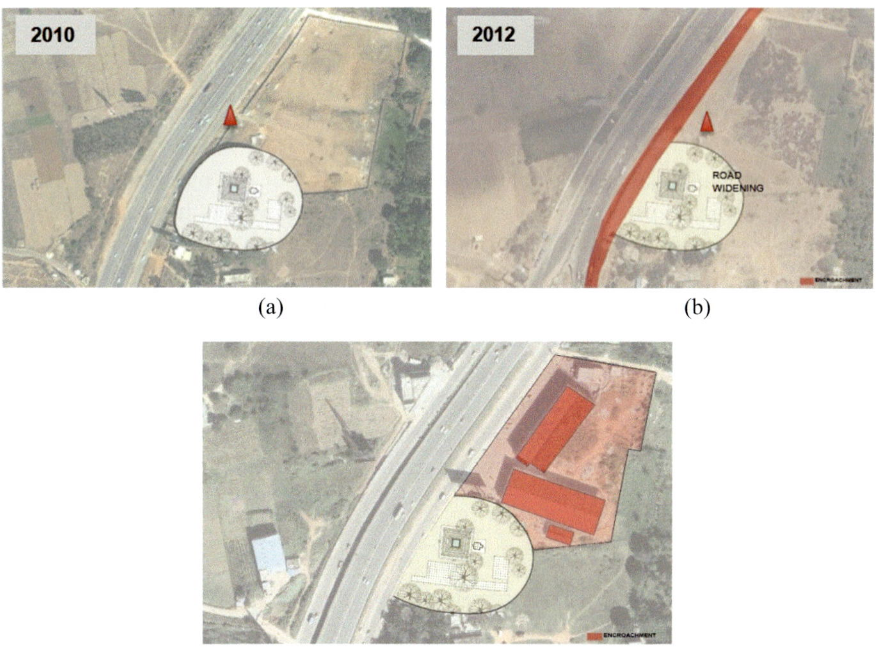

Fig. 18 Changes through time—**a** Enclosed fortification, 2010, **b** Road widening, 2012, and **c** Encroached construction, 2022. *Source* Author

the Archaeological Survey of India (ASI), it has been removed from the conservation list. In the year 2017, there was a proposal for de-notification as there was a dispute that had emerged with the owners. The owners sold the land to private construction companies for their purposes. In such situations, the monument under national importance can be ceased as per clause 35 of the AMASR Act, 1958; the act states that: If the Central Government is of the opinion that any ancient and historical monument or archaeological site and remains declared to be of national importance by or under this Act has ceased to be of national importance, it may, by notification in the Official Gazette, declare that the ancient and historical monument or archaeological site and remains, as the case may be, has ceased to be of national importance for this Act. Through this, the heritage structure is no longer under the ASI.

From the figure (Fig. 19), it can be seen that the Chikkajala fort has been under negligence for a long time. Excess growth of vegetation has caused a lot of structural damage. The main affected area is the *Dharmachattra* (pillared hall) (Fig. 19A, B). It has a lot of structural cracks and vegetation growth, wherein a tree has grown which penetrated through the roof of the hall causing major issues to its structural stability making it difficult for its revival.

The *Kalyani* (Fig. 19C) is dysfunctional and does not contain water anymore. There is a lot of vegetation growth. The front part of the fort wall has been demolished and now it doesn't exist giving easy access to all kinds of vandalism (Fig. 19D). At present, the Chikkajala fort is used by localities and auto drivers for resting (Fig. 20).

6 Discussion

The study of Chikkajala Fort, located on the outskirts of the metropolitan city of Bengaluru, emphasizes establishing the significance and values it possesses. From the study, it can be inferred that the fort complex possesses historic, cultural, architectural, and socio-economic values. Urbanization has created problems for the Chikkajala fort complex, and an ASI-managed prehistoric site has been eliminated as a result of ongoing changes and developments.

The culture is fading because the structures are not being adequately preserved because of several recent developments. The values that are associated with the fort complex make it significant to date. These values have been associated with its times dating back primarily from the Hoyasala period (eleventh to fourteenth century) and therefore need attention.

7 Conclusion

The neglect of the Chikkajala fort complex caused due to rapid urbanization and growth of the city, as well as the lack of protection from any competent authority, has resulted in its dilapidation. Hence, there is a need for the government to create

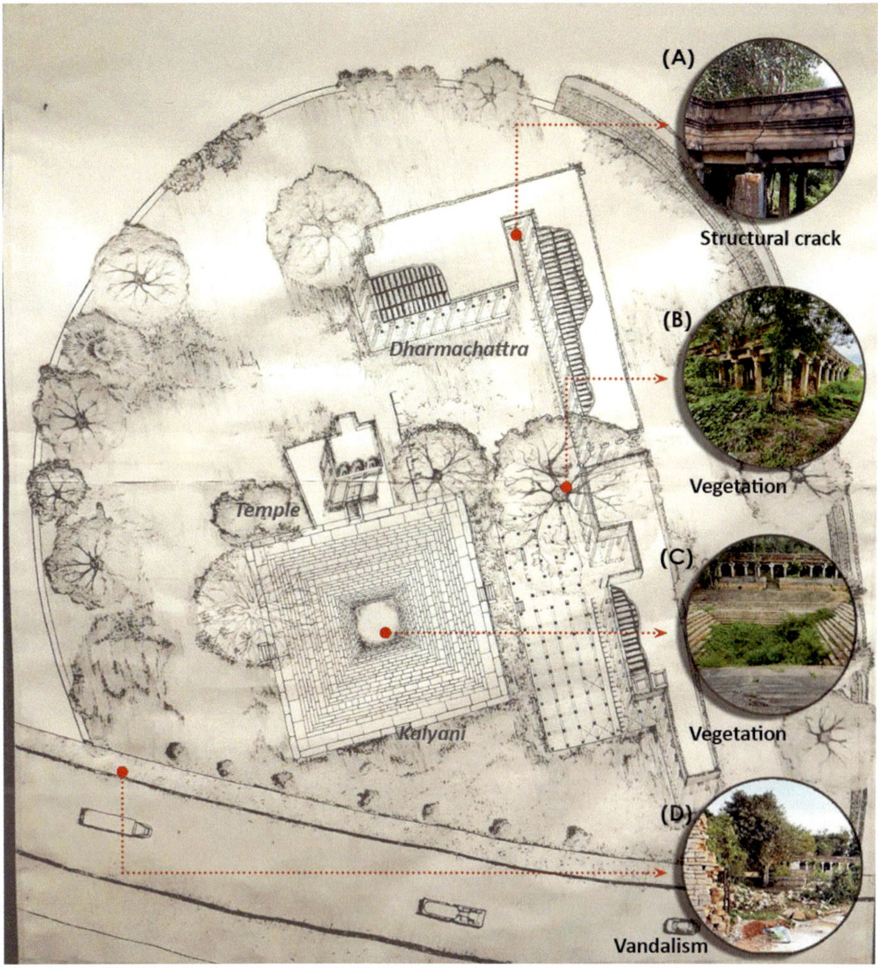

Fig. 19 Present condition. *Source* Author

stringent policies that must be followed for the protection of its heritage in the city. These policies must be incorporated at the master plan level itself. As suggested by SDG 11.4, cultural rights specifically are often exercised at the local level 2 accordingly; local governments have a very significant role in the design and implementation of cultural policies.

The existing proposal for a light and sound show which was by the Karnataka State Tourism Department in the year 2010 can be adapted as a proposal and can be implemented to boost tourism as well. The rapid urbanization in the area can be taken advantage of by creating public awareness to boost tourism for the Chikkajala fort. The revenue generated from the tourism strategy could be used for the regular maintenance of the fort itself.

Fig. 20 Resting zone under a Peepal. *Source* Author

References

Abhishanker K (1990) Karnataka State Gazetteer: Bangalore District. Publisher, State Government of Karnataka

Bhargava P (2010) Hoyasala Temple Architecture—The Later Chalukyan Style. In: Proceedings of the Indian History Congress 71:1307–1309. http://www.jstor.org/stable/44147623

Chikkajala Fort (2015). https://www.thetravelreminiscences.in/2015/09/chikkajala-fort.html

Gogoi VAPBA (2018) ASI to de-notify Chikkajala megalithic site. Angarikagogoi. https://angarikagogoi.wordpress.com/2017/09/14/asi-to-de-notify-chikkajala-megalithic-site/

Gupta E, Das S, Balan KSC, Kumar V, Rajani MB (2017) The need for a National Archaeological database. Curr Sci 1961–1973

Jagannath U (2011, December 31) NHAI demolishing historic Chikkajala fort for road widening. Citizen Matters, Bengaluru. https://bengaluru.citizenmatters.in/3678-nhai-demolishing-historic-chikkajala-fort-for-road-widening-3678

Kirana Bhat (2020) https://kiranabhat.com/2020/10/25/chikkajala-fort-or-not/

Nast C, C. (n.d.). Condé Nast Traveller India. Condé Nast Traveller India | the Latest Travel Updates and Advice for the Indian Traveller | Condé Nast Traveller India. https://www.cntraveller.in/

Priya (2020) https://www.cntraveller.in/story/bengaluru-hiding-in-plain-sight-a-3000-year-old-relic-chikkajala-fort/

Vijayalaxmi J, Arathy Kalam KC (2022) Critical evaluation of socio-cultural and climatic aspects in traditional weavers' community—a case of Pillayarpalayam weavers' cluster, Kanchipuram. Built Heritage 6(1). ISSN: 2662-6802

Study of the Planning and Architectural Significance of the Chitnavis Wada in Nagpur, India

Shreya S. Kalbhor

Abstract The traditional dwellings in Maharashtra are called *Wadas*. The royal families in Maharashtra during the period of Bhosalas in Nagpur planned and designed these Wadas as their residences back in the seventeenth century. Chitnavis Wada in Nagpur is one of them which served as a residence for the ministers of Bhosale in the eighteenth century. Since the descendants of these Wadas have moved out for better prospects, the physical structure is in danger of dilapidation. This historic structure is either getting vandalized or they are being reused as a museum or gallery to increase the economic graph. Due to this, a loss of historicity can be observed in the Wadas, which are originally meant to serve as residences for the royal families. There is a need to conserve this piece of history, which reflects the socio-cultural and economic times of 18th-century dwellings of the royal families. The aim of this paper is to understand the planning and architectural significance of the Chitnavis Wada complex in Nagpur. The study is based on the on-field documentation of the Chitnavis Wada in Nagpur. The identified factors will further be identified to understand its architectural significance.

Keywords Architectural significance · Dwellings · Bhosale · Wada · Royal families

1 Introduction

The architecture of Maharashtra dates to the seventeenth and eighteenth centuries with high architectural and cultural significance. During the sixteenth and seventeenth centuries, after the establishment of the power of Bhosale in Nagpur, several small, medium, and magnificent scale traditional dwellings of a particular style came into existence. The distinctive type of traditional housing in Maharashtra is "Wada".

S. S. Kalbhor (✉)
School of Planning and Architecture, Vijayawada, Vijayawada, India
e-mail: kshreya645@gmail.com

Vijayalaxmi J. (ed.), *Conservation of Built Heritage in India*, Advances in 21st Century Human Settlements, https://doi.org/10.1007/978-981-99-8335-3_4

Generated during the period 1700–1900. Maratha style of architecture, in particular, is influenced by various economic, political, cultural aspects, and topographical conditions.

The character of housing types in India has undergone many changes since the eighteen century. The housing types have changed entirely, from the exteriors to the planning. Changes in the character of housing are the outcome of different kinds of people (the migrants from other states), their lifestyles and their societies. In due course of urbanization, the traditional structure of families changed to nuclear families. Only some courtyard houses survive as a result of the dismantling of the feudal system. The structures that have survived were built as palaces for kings and the elite.

Many authors have studied the courtyard evolution and the impact of urbanization on the Wadas in Maharashtra (Alapure et al. 2016; Kukdolkar 2016; Deshpande et al. 2015). There is less understanding of how to preserve this piece of history by considering the planning and architectural significance of these traditional dwellings of Maharashtra. Hence the study aims to understand the planning and architectural significance of the Chitnavis Wada in Nagpur. The study is based on the documentation of the Chitnavis Wada. The strategies for the adaptive reuse of the Wada are enlisted to ensure their significance. Chitnavis Wada consists of the old Murlidhar Mandir dating back to the eighteenth century, which has its unique architectural, historical, and socio-cultural value, marking its importance within the Chitnavis Wada complex and creating an opportunity for its conservation. As suggested by SDG 11,there is a need for bigger efforts to protect the world's cultural and natural heritage. As a result, this paper examines the condition mapping of the Murlidhar Mandir and suggests the primary level of interventions for the same.

2 What is Wada?

The traditional residence in Maharashtra was called the Wada (Fig. 1). A Wada was typically a large building of two or more storeys with groups of rooms arranged around open courtyards. Its style was an amalgamation of features from Mughal, Rajasthan, and Gujarat architecture combined with local construction techniques. Wada is a kind of introverted house form and thus Court becomes a point of emphasis. It reflects the rich culture, tradition and pride of the turbulent history of the Marathas Gogate (2015).

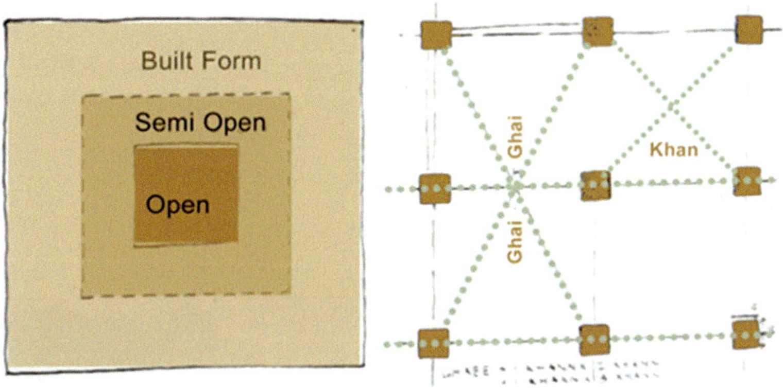

Fig. 1 Conceptual plan of Wada and its structural grid expansion (*Source* Author)

3 Types of Wadas

i. Garhi Wada- The Garhi or rajwada being a fortified structure was introverted and self-sufficient with ramparts and bastions as protection against invaders.
ii. Noble Men's Wada- It belonged to the noblemen of the court of Peshwas. The internal spaces are planned according to the daily activities of the noble Peshwas. It lacks fortification, suggesting stable and peaceful times during the Peshwa period, with a focus on more ornamentation.
iii. Ordinary Wada- The ordinary Wada belongs to common people who existed across all regions. It usually has a single courtyard and an open yard at the back.

4 Study Area

4.1 Location and Climate

Nagpur is located in the northeastern region of Maharashtra i.e., Vidarbha region (Fig. 2). Chitnavis Wada is located in the heart of old Nagpur in the Chitnavis area. In Nagpur, the monsoon season is oppressive and overcast, the summer season is mostly clear, and it is hot year-round. Over the course of the year, the temperature typically varies from 56 to 108°F and is rarely below 50°F or above 113°F.

4.2 History of the Chitnavis Family

Chitnavis Wada has a strong historical significance in Bhosale's history as they served as a Chitnavis (political secretary) to Bhosale. Chitnavis followed the Bhosales from

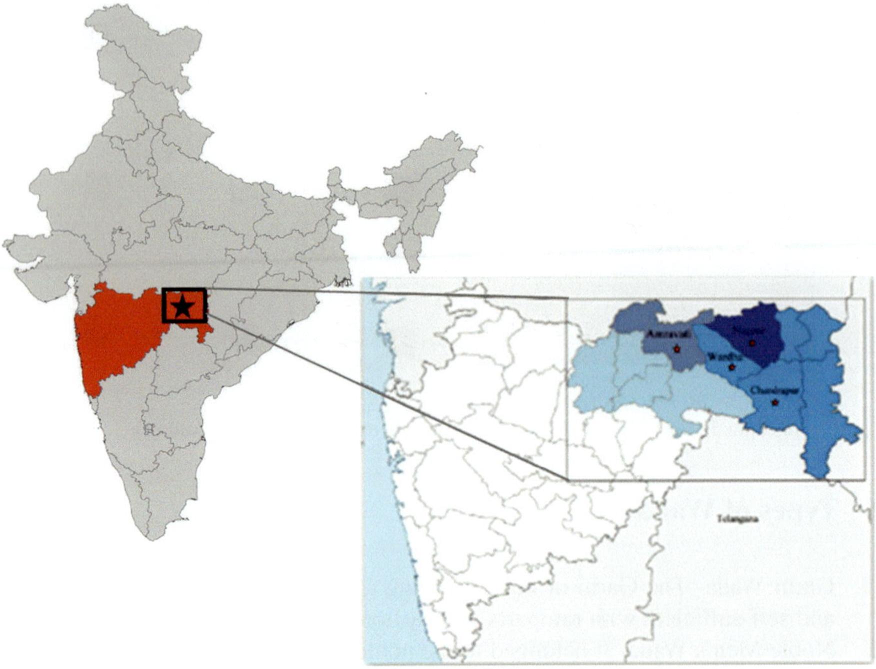

Fig. 2 Map of India showing the Vidarbha region, Maharashtra (*Source* Baig et al. (2017a, b))

Berar to Nagpur in 1743 AD with Raghuji Bhosale I. The Wada was built in 1760 AD (Heritage Conservation Committee 2015).

Many generations served the Bhosale as ministers (Chitnavis) and secretaries at the court of Nagpur until the middle of the nineteen century. Gangadhar Rao Chitnavis was president of the Nagpur Municipality (Kumar n.d.).

Displaying a language or rich tradition, the Wada is a Grade I structure on a site area of around 8000 sq. m. It has a large verandah and walls covered with exquisite artwork and a rich display of decorated timber pillars (Mathur 2016).

4.3 Phases of Construction

Chitnavis Wada complex consists of 2 phases of construction i.e., old Chitnavis Wada which was built in the mid-eighteenth century by the 1st Chitnavis and the extension part of the Wada was built early twentieth century by Gangadhar Rao Chitnavis. The complex consists of Murlidhar Mandir which dates back to the mid-eighteenth century (Fig. 3).

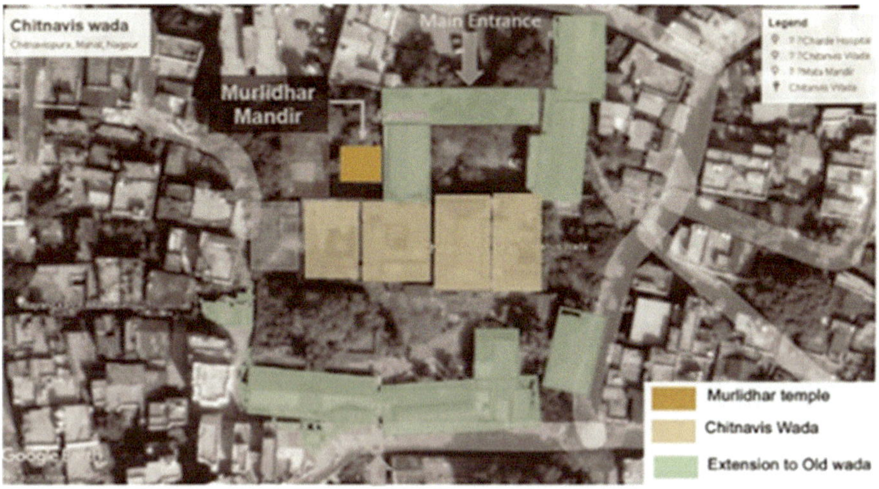

Fig. 3 Satellite image showing the zoning of Chitnavis Wada complex (*Source* Google earth)

4.3.1 Phase 1 (1750–1866)

In phase 1 Rakhamji Randive (1st Chitnavis) constructed the main Wada with three courtyards, Murlidhar mandir at the north surrounded with servants' quarters and Paga (space to park horses) at the southern side as shown in Fig. 4.

The Central Core and Its Spatial Planning

The various spaces in the Chitnavis Wada are detailed in Fig. 5.

Osari: It is the transition space or verandah. It is a semi-open space or a passage or spill-out space for activities.
Dewadi: A verandah for guards
Sadrecha Sopa: Open to the courtyard, the verandah space was used for administrative activities, usually in the first or central courtyard.
Kacheri: Office: It is the administrative department in the first or central courtyard.
Diwankhana: Living room. Huge hall for formal meetings.
Majghar: Middle room. From this part, the private area is segregated from the public area. Generally, women & family members use it. It is a private hall
Devghar: Prayer Room
Tijory: Treasury.
Gotha: Cow-pen in the backyard of a house.
Kothar: Store room.
Swayampak Ghar: Kitchen

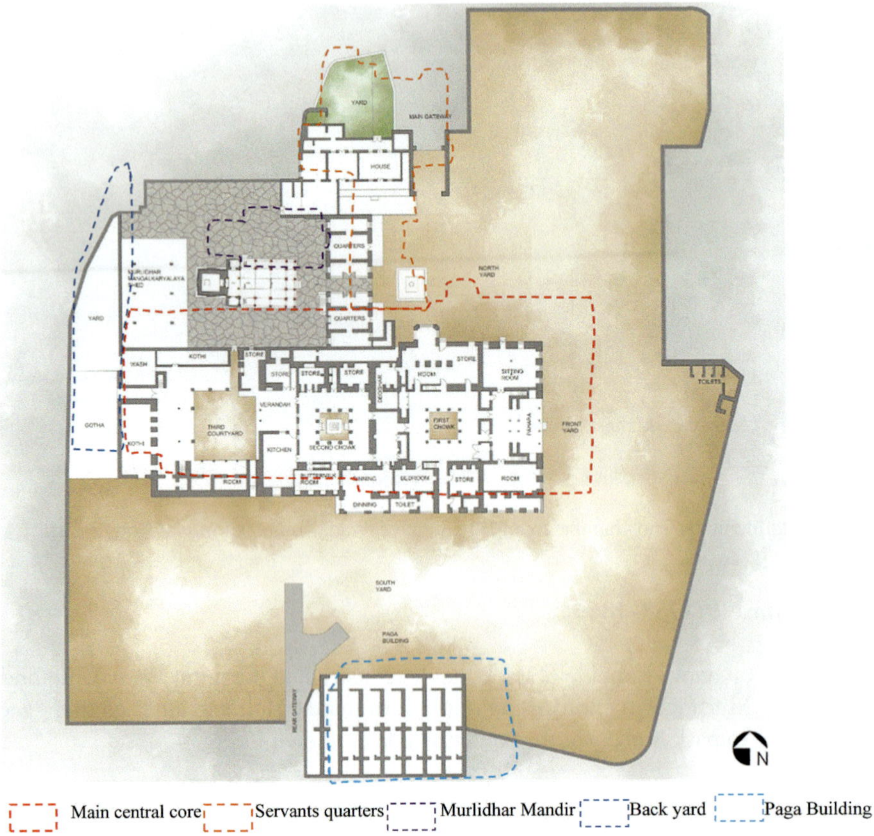

Fig. 4 Plan showing the Chitnavis Wada in 1st phase, eighteenth century (*Source* Author)

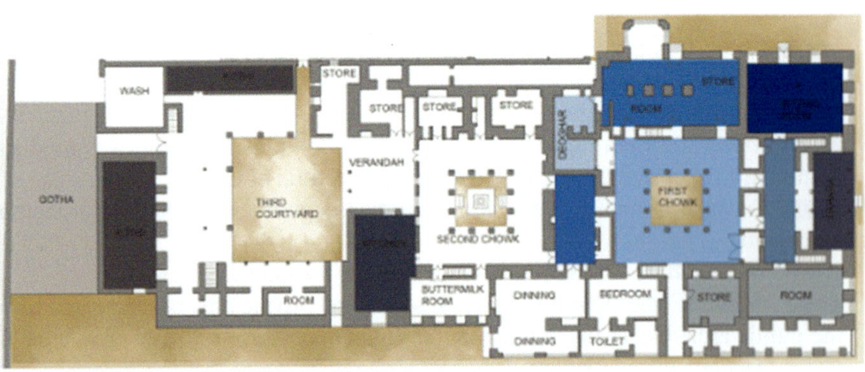

Fig. 5 Plan showing the spatial layout inside the old Chitnavis Wada (*Source* Author)

Fig. 6 2nd courtyard

Fig. 7 3rd courtyard

Use of the Courtyards

The first courtyard acts as a semi-private space dedicated to carrying out administrative meetings and discussions. The second courtyard acts as a private space for the family get-together and as a common dining area for the family (Fig. 6). The third courtyard is dedicated to women of the houses to perform their daily rituals and household activities which is completely private (Figs. 7 and 8).

Architectural Features

There are wooden windows and doors used in Wada. Simple wooden and door frames have been used for both windows and doors with lower lintel heights. Sagwan wood is used for all the wooden construction in Wada (Fig. 9). All the staircases are built with adjoining 1.2 m thick walls. The staircases were narrow in width with just 80 cm wide treads (Fig. 10). This way the privacy of the house was maintained. All

Fig. 8 3rd courtyard (*Source* Author)

Fig. 9 Types of doors

the external walls of the Wada were 1.2 m thick. This helped to keep the interior of the wall cool in the summer.

The niches provided in the Wada were a combination of both Hindu and Mughal architecture. They are known as *Dewali* in Hindu architecture. It was made to place lamps inside them as they protect the light from the wind (Fig. 11).

Wooden bracketed columns (Fig. 12) with elaborate ornamentation of brackets (Fig. 13) have been used.

A detail showing the column and brackets attached to it bearing the load of the internal overhang of the courtyard ceiling is shown in Fig. 14.

Fig. 10 Staircase

Fig. 11 Dewali (*Source* Author)

Construction Details

Wooden beams carry the load of the structure and pass it in the column through brackets, which in turn transfer it to the ground through the column and other ends of the beam are supported by load bearing wall. Wooden boards are placed on a wooden joist for wood on which mud or stone floor is made with lime mortar (Fig. 15).

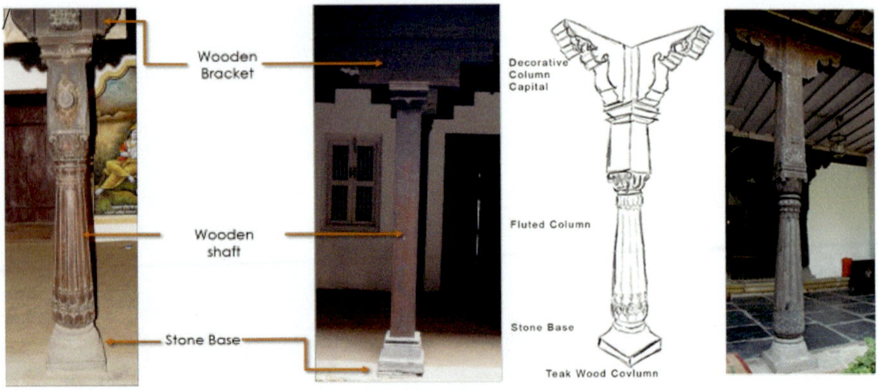

Fig. 12 Column details in the courtyards (*Source* Author)

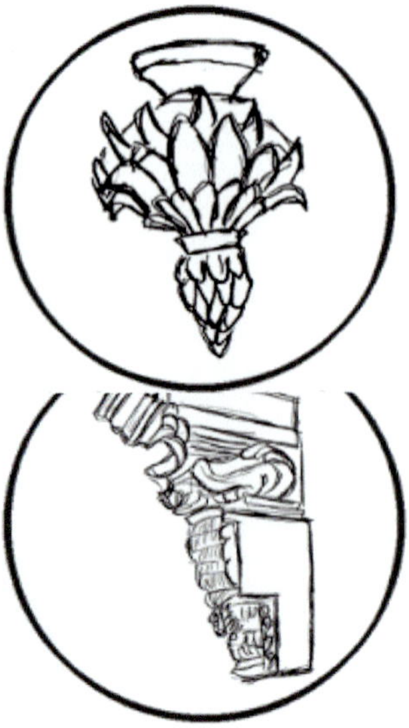

Fig. 13 Bracket ornamentation details (*Source* Author)

Fig. 14 View showing brackets supporting ceiling joists and detail showing column and bracket attached to it bearing the load of the internal overhang of the courtyard ceiling (*Source* Author)

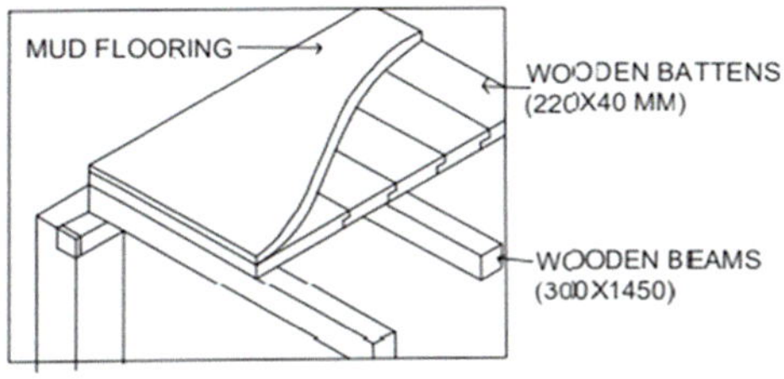

Fig. 15 Floor detail (*Source* Author)

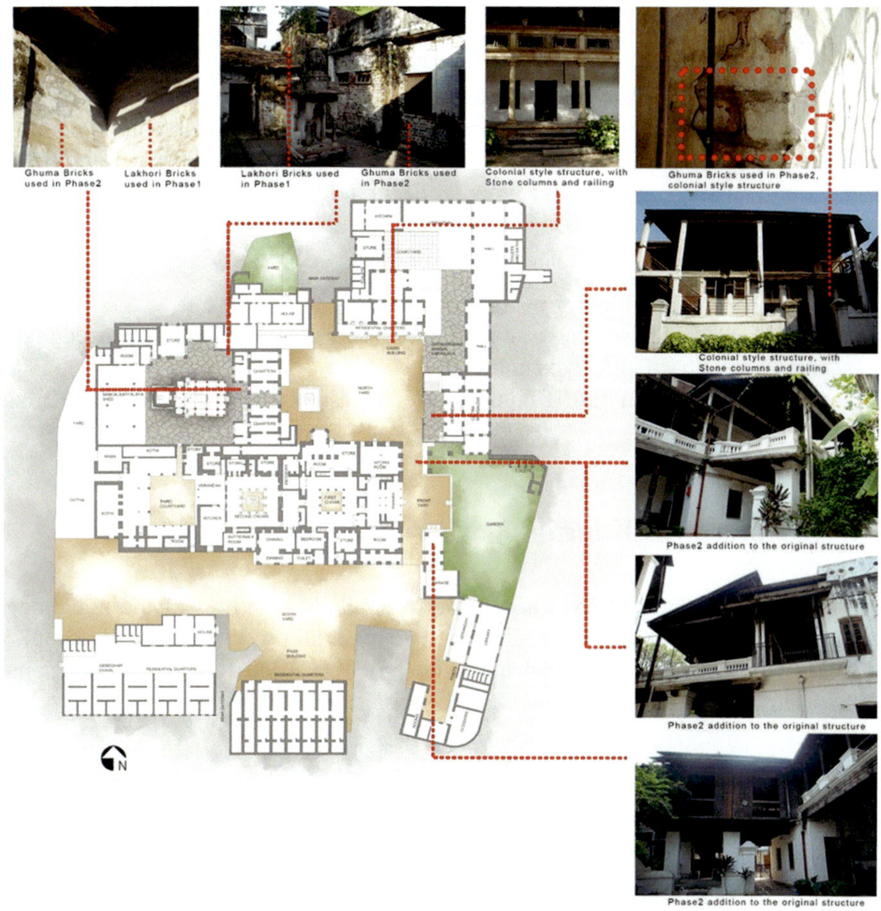

Fig. 16 Plan showing the additions made in the 2nd phase, early twentieth century (Fig. 16) (*Source* Author)

Material Mapping

The materials which have been used in the construction of phase are primarily Sagwan wood, sandstone, black basalt stone and lime plaster.

4.3.2 Phase 2 (1866–1940)

Gangadhar Rao was appointed as a member of the Imperial Legislative Council in 1904 and represented the Landholders Constituency from the Central Provinces and Berar for several years. (Built new structures around Wada, part of which was Britishers Guest House).

During the second phase, we can see the addition of new materials like *ghuma* bricks for the construction. The use of materials and the typical architectural features infers the colonial period. The use of ghuma bricks for the construction, round stone columns with capitals, and wooden decorated railings indicate the colonial architectural features (Fig. 23).

Segregation of phase 1 and phase 2 is based on:

(i) Wall Thickness: which varies in both phases.
(ii) Column Style: Phase 1 has a stone-based timber column with intricate capitals, and Phase 2 has Doric-type columns.
(iii) Column Material: Phase 1 columns are consistently timber, and phase 2 has stone columns.

4.4 Architectural Value of the Chitnavis Wada

The Chitnavis Wada is planned according to the daily activities of the noblemen, having office spaces inside the residence. The material used in the Wada complex are indigenous and the planning and zoning of the Wada follow the traditional Maharashtrian style. Ornamentation and details are true to their time.

4.5 Historical Value of the Chitnavis Wada

Chitnavis Wada is the only Wada complex remaining in Nagpur having a temple within its complex. It belongs to the Chitnavis family which was one of the royal families in the court of Bhosale in Nagpur, Chitnavis served as minister for the Bhosale. The Chitnavis Wada is an example of rich Maratha heritage.

5 Materials and Methods

The methodology includes literature review, on-field documentation, condition mapping, historical and architectural analysis, and assessment of intangible factors. Interviews and surveys were conducted to gauge visitor perceptions and awareness. The research highlights the unique features, construction techniques, and association with the Chitnavis family and the Bhosale dynasty. It emphasizes the importance of preserving cultural heritage amidst urbanization. The study proposes conservation measures, including adaptive reuse of certain portions, to ensure the sustainability of Chitnavis Wada, protect its historical significance, promote tourism, and provide economic opportunities for the local community.

5.1 Tools and Techniques Used for the Documentation

The various tools used for physical documentation of the Wada are shown in Fig. 17.

The techniques to map the physical documentation are shown in Figs. 18, 19, 20, 21 and 22.

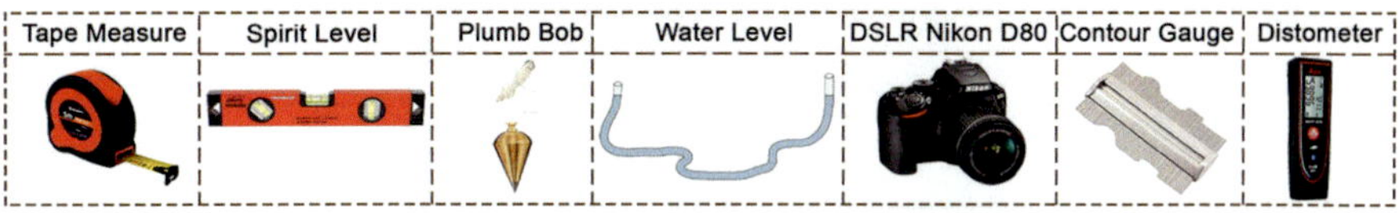

Fig. 17 Tools of primary documentation (*Source* Author)

Fig. 18 Marking of the plan cut level with the help of water tube on the tricky surfaces. (*Source* Author)

Fig. 19 Marking of the plan cut level with the help of water spirit on the flat surfaces (*Source* Author)

Fig. 20 Contour gauge to get column profile in plan. (*Source* Author)

Fig. 21 Marking on columns at cut level with the help of a water tube and using of water tube at the extreme corners of the structure to check the slope (*Source* Author)

Fig. 22 Use of windows of neighboring structure to achieve photogrammetry for temple shikhara. (*Source* Author)

6 Case of Murlidhar Mandir (Phase 1) for the Restoration Proposal

6.1 Murlidhar Mandir

Murlidhar Mandir was built in Phase 1. Even though there is Deoghar present in the 1st courtyard of Wada, the Chitnavis family had built a separate temple dedicated to lord Murlidhar to establish their status symbol.

The temple is dedicated to lord Murlidhar, Rukhmini and Rai. The idols are made of white marble. The temple has two smaller shrines on both sides of *Antarala* dedicated to lord Shiva and lord Ganapati. Both idols are made of black basalt.

6.2 Architectural Value of Murlidhar Mandir

Murlidhar mandir has its unique style which is an amalgamation of different architectural styles. The open timber mandapa reflects the impact of Maratha architecture during the Bhosale rule.

6.3 Historical Value of Murlidhar Mandir

The women of the house being of higher status needed a lot of security to go out of the house to ensure their safety. To avoid such situations and for the ease of the family, people of noble families used to build private temples inside the Chitnavis Wada complex. Every family has a family god and goddess called *kuladaivat* and *kuladevi* from a certain temple which is not always present in the same city. Travelling such long distances to visit the temple was not always feasible, thus Nobel families like Chitnavis and Bhosale in Nagpur used to build the temples of their kuldaivat or kuldevi inside the Wada complexes.

6.4 Socio-cultural Value of Murlidhar Mandir

Various events and functions are conducted in the temple some of which are for lord Murlidhar, such as Krushna Janma or Gokulashthami and Gopalkala. Family *Havans* are also conducted in the temple. Havan is a ritual wherein offerings such as rice and clarified butter are made into a consecrated fire. Presently, the Wada complex is rented for various functions out of which marriage rituals and havans are conducted in Murlidhar Mandir. On a daily basis, a priest appointed by the Chitnavis family opens the sanctum sanctorum (*Gabhara*) at 8 am and 7 pm for daily prayer rituals (*Pooja*).

6.5 Condition Mapping and Analysis of Murlidhar Mandir

The condition mapping of the Murlidhar mandir is shown with the help of plans (Fig. 23), sections (Figs. 24 and 25), and elevation (Fig. 26) respectively. The comprehensive listing of the condition is presented in Table 1.

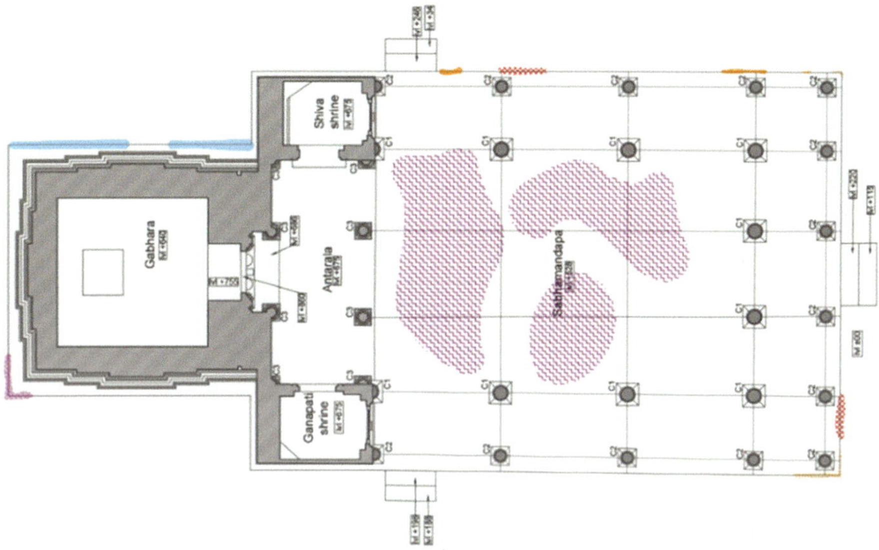

Fig. 23 Plan of Murlidhar temple (*Source* Author)

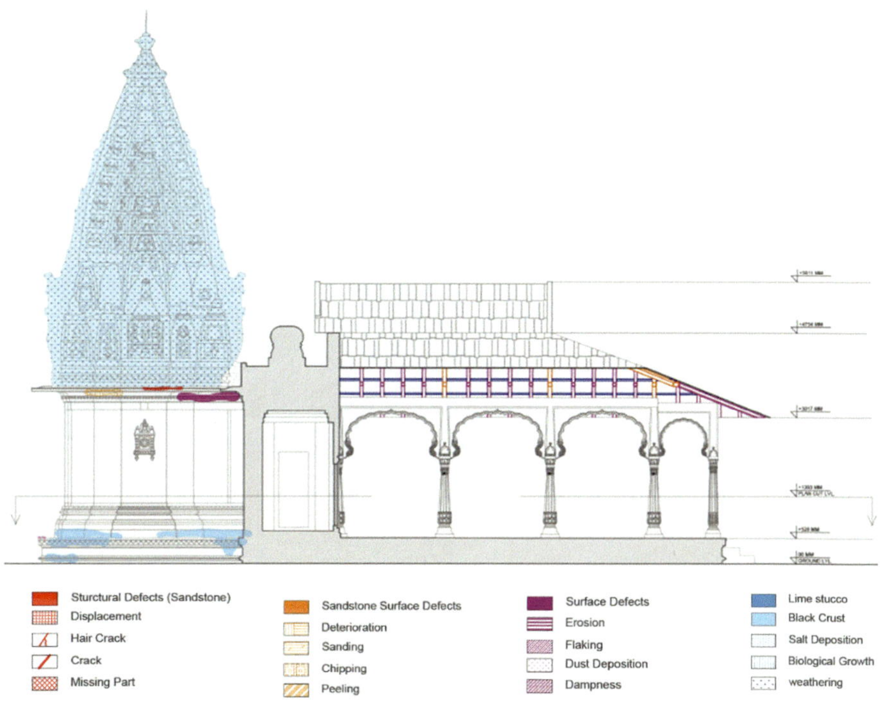

Fig. 24 Longitudinal Section through the Temple (*Source* Author)

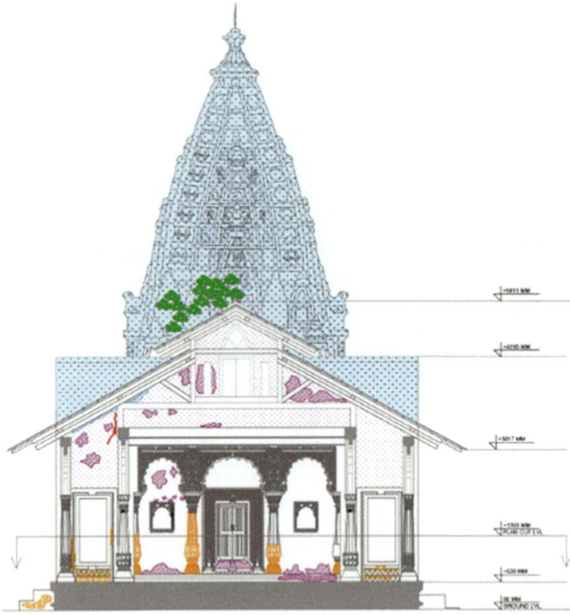

Fig. 25 Lateral Section of the Temple

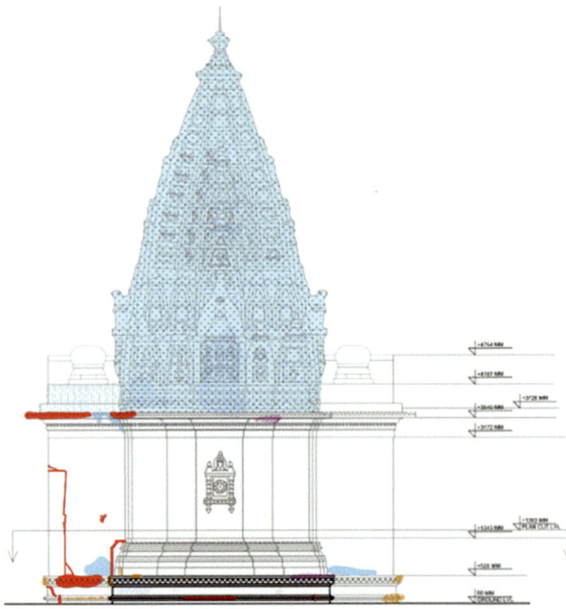

Fig. 26 Elevation- West (*Source* Author)

Table 1 Table showing the defects and primary interventions

Defects	Pictures	Interventions
Displacement-Overloading of excessive load over 70 mm stone wall		For displacement of stone, the addition of stone masonry adjoining to the old 70 mm existing masonry. Filling the cracks with lime mortar. As missing parts are just decorative and not structural it needs to be further maintained
Hair Crack-Expansion and contraction in the sandstone		
Fracture-Expansion and contraction in the sandstone		
Missing Part-Negligence		
Erosion-Exposure to climatic conditions		Re-application of the lime plaster by using traditional ingredients. General upkeep of the structure
Flaking-Negligence and exposure to climatic conditions		
Dust Deposition- Negligence		
Dampness-Rising water from ground to plinth		
Negligence and exposure to climatic conditions		
Biological Growth-Water penetration and exposure to moisture for a longer time		Algae removal by warm water, soft nylon brush, and wet tissue paper. Try to eliminate the causes of water penetration and regular maintenance. Cleaning the salt deposition using the paper pulp method
Black Crust-Exposure to moisture for a longer time		
Salt Deposition-Exposure to Moisture for a longer time		

Source Author

6.6 Intangible Aspects Affecting the Condition of the Structure

In the years that the Chitnavis used to live in the Wada, the Wada was opened for the general public only on special occasions like Gokulashtami and Ganeshjanma (which are important Hindu festivals). But in the present day, even though the Wada and temple are open to all, people still follow the same practice of visiting the temple only on special occasions. The main access to the temple from the north side is mostly always kept closed. Hence people visiting the Wada can only access the temple from the 3rd courtyard. Due to this inconvenience of access to the temple, many are not even aware of its existence. The temple is lesser known, as people visit the complex with the intention of visiting the residence and later realize the existence of the temple after the entire tour of Wada.

7 Proposal

7.1 Conservation Proposal for the Chitnavis Wada Complex

In order to ensure that the Chitnavis Wada does not get degraded over time, it is essential to reassess the appropriate conservation tools to ensure its sustainability. Adaptive reuse in the form of alternative spatial provision for the local woodwork craftsmen in one part of the complex can bring in tourists. In this manner, the historical significance of the building as well as the craft can be sustained. The movement of the public in the Chitnavis Wada is conceptualized in Fig. 27.

7.2 Adaptive Reuse for Phase 2 (Chitra Bungalow)

The Chitra Bungalow from phase 2 has been proposed for adaptive reuse as a weaving center (Fig. 28). During the rule of Bhosale in Nagpur, the ladies of the Bhosale family and noblemen's family including Chitnavis used to drape the *Kashta* Saree (Fig. 29) woven by the Muslim weavers of Mominpura, an area at a distance of 750 m from the Chitra bungalow. At that time this traditional craft of making Kashta saree flourished due to its recognition by Bhosale's in Nagpur.

This weavers' community at Mominpura was established during the Bhosale rule by Raghuji Bhosale II and belonged to the Muslim community of weavers who were engaged in the handloom business. The weavers of the Mominpura area of Nagpur have been engaged in their traditional handloom craft of making Kashta sarees since the period of Raguji Bhosale II.

Weavers living in the Mominpura area, are operating this business in their houses with the help of traditional looms because of the shortage of proper infrastructure. Due to the lack of proper training, youth from this community are shifting to other occupations. Thus, the traditional knowledge system of weaving this drape is on the verge of extinction. By introducing the weaving handloom in the Chitra bungalow, the dying weaves as well as the heritage of the Chitra bungalow can be revived. The vernacular architecture of weavers' communities represents one such unique architectural typology, where the built environment is shaped by their craft and must be preserved (Vijayalaxmi and Arathy 2022).

8 Discussion and Conclusion

The Chitnavis Wada has an architectural, historical and functional value from the past, which lies unexplored. The Wada is an important piece of history of its times. This study has documented the Wada in detail and has proposed measures to ensure

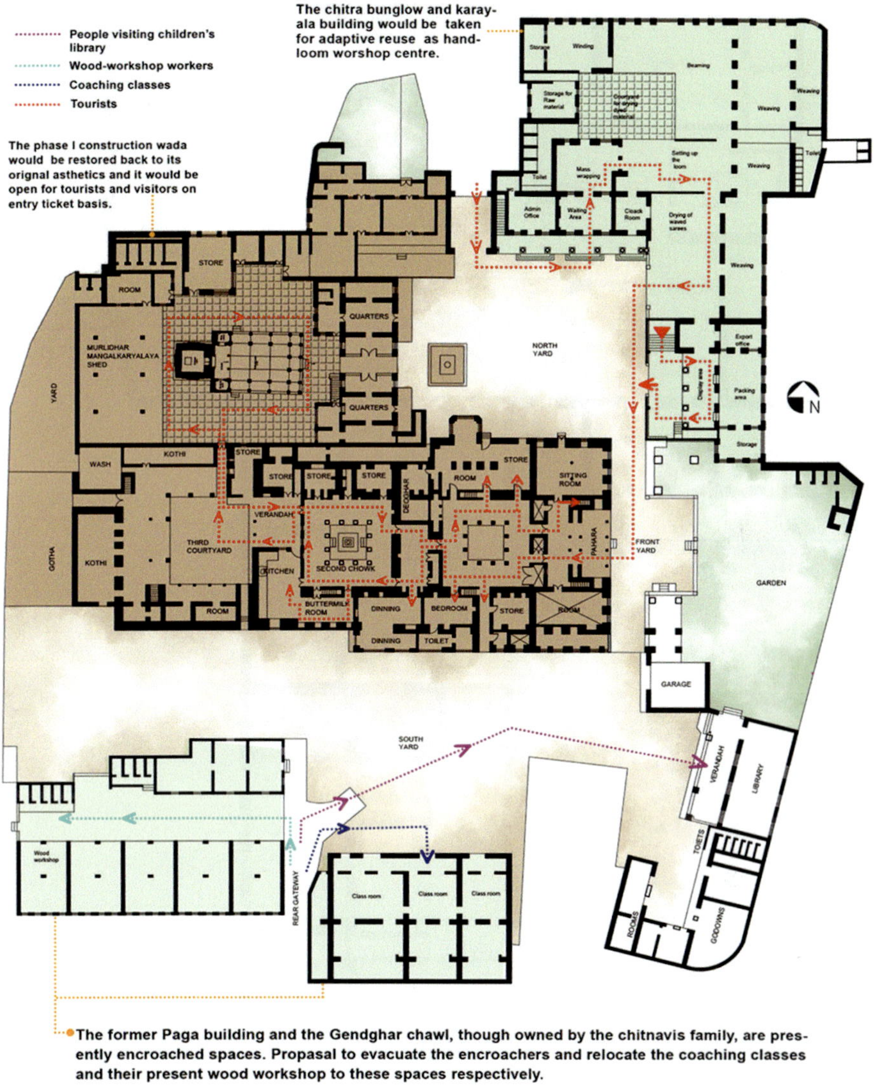

Fig. 27 Plan Showing the proposed circulation for the complex and proposed interventions. (*Source* Author)

that the building does not die a silent death, due to a lack of knowledge of its presence as well as relevance.

Restoration has been proposed for the Wada to regain its original aesthetics. For the 1st phase of the Wada, it is proposed to promote tourism and create awareness among the public about the significance of the structure.

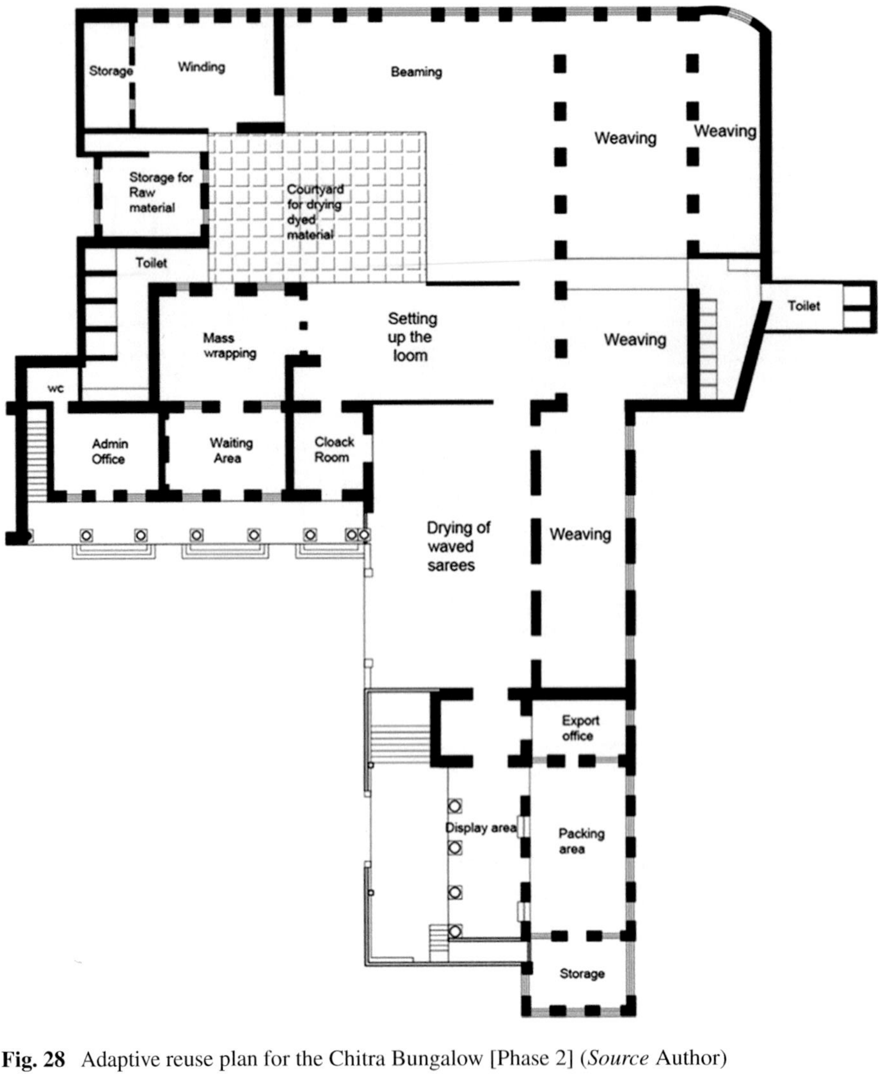

Fig. 28 Adaptive reuse plan for the Chitra Bungalow [Phase 2] (*Source* Author)

Adaptive Reuse is proposed for the 2nd phase of the Wada for better spatial usage and also to prevent encroachment by outsiders. A weaving center is proposed for the local weavers of the Mominpura area, thus conserving and promoting the traditional knowledge system of the weaving community as well as ensuring the use of the bungalow. This would in turn develop the economic value of the community. Conservation measures have been proposed for the Chitnavis Wada so that the aesthetics are restored back and they generate better incomes, offer job opportunities for the local community and also ensure that the traditional knowledge system is preserved for generations to come. As suggested by SDG 11, in the context of urbanization,

Fig. 29 The Kashta saree drape 2023 (*Source* Drapes of India)

which puts pressure on heritage assets, it is important for cities to protect cultural heritage and place culture as a key aspect in the renovation or historic centers and in inclusive urban planning.

It is important for such places of historic importance to be conserved, as their dilapidation and loss could mean a loss of a precious piece of history.

References

Alapure GM, Bhattacharya SP, George A (2017) Climate responsiveness of Wada architecture. In: Seta F, Biswas A, Khare A, Sen J (eds) Understanding built environment. Springer transactions in civil and environmental engineering. Springer, Singapore, pp 29–25. https://doi.org/10.1007/978-981-10-2138-1_3

Baig M, Mahakalkar A, Sapkal H (2017a) Map of India showing the position of Eastern Maharashtra (Vidarbha) [Map]. Report of high genetic diversity of filarial worm, Wuchereria bancrofti from endemic region of Eastern Maharashtra (India). Helminthologia 54(4). https://doi.org/10.1515/helm-2017-0043. https://www.researchgate.net/

Baig M, Mahakalkar A, Sapkal H (2017b) Map of India showing the position of Eastern Maharashtra (Vidarbha) [Map]. Report of high genetic diversity of filarial worm, Wuchereria bancrofti from endemic region of Eastern Maharashtra (India). Helminthologia 54(4). https://doi.org/10.1515/helm-2017-0043. https://www.researchgate.net

Deshpande R, Kotharkar R (2015) ‘Dwellings’ then and now: a topological approach for privacy analysis of ‘Wada’ and modern houses. In: Proceedings of 10th international space syntax conference, 2015. https://www.researchgate.net/

Drapes of India—The Maharashtrian Nauvari saree (2023) Sundari Silks. https://www.sundarisilks.com/blogs/article/drapes-of-india-the-maharashtrian-nauvari-saree

Gogate RV (2015) Peshwyanchi Bakhar: a study of housing typology —Wadas—in historic urban cores of Pune and Nashik. Post Graduate Thesis. CEPT, Ahmedabad. http://hdl.handle.net/20.500.12725/6709

Heritage Conservation Committee (2015) Heritage Catalogue of Nagpur. Nagpur Municipal Corporation. https://www.nmcnagpur.gov.in/assets/250/2021/02/mediafiles/9-138.pdf

Kukdolkar P (2016) Revitalization of traditional housing: a case of Wadas in old Pune. Post Graduate Thesis. CEPT, Ahmedabad. http://hdl.handle.net/20.500.12725/14174

Kumar S (n.d.) The architecture of Wadas of Maharashtra. rethinking the future. https://www.re-thinkingthefuture.com/rtf-fresh-perspectives/a828-the-architecture-of-Wadas-of-maharashtra/The%20Architecture%20ofWadas%20of%20Maharshtra%20as%20read%20on%2020th%20Dec.%202022

Mathur B (2016) 250 years ago once again. The Times of India. https://timesofindia.indiatimes.com/city/nagpur/250-years-ago-once-again/articleshow/52980727.cms

Vijayalaxmi J, Arathy Kalam KC (2022) Critical evaluation of socio-cultural and climatic aspects in traditional weavers’ community—a case of Pillayarpalayam weavers’ cluster, Kanchipuram, Built Heritage, volume 6, issue 1, ISSN 2662-6802

A Study on Defense Planning Strategies of Gwalior Fort

Muskan Tiwari

Abstract During the reign of the Tomar Rajputs in the twelfth century, the surrounding kingdoms were trying to expand themselves by strengthening their defense systems. The biggest threat which the kings were facing was from the Mughal invasions. Kings were trying to build forts with in-built defense techniques. At that time, one of the forts flourished amongst the most impenetrable mountain forts in India, Gwalior fort located in the Madhya Pradesh state of India. In order to appreciate the planning of Gwalior fort there is a need to understand their defense technique and its implication in the planning of Gwalior fort. This study aims to assess the defense architecture technique of Gwalior fort through archival records supported by field studies and primary information from government agencies and historians. The study shows that the northeast part of the Gwalior fort has been planned primarily to prevent the fort from the Mughals because the maximum attacks that happened in the fort were from the northeast gate. This study will help in appreciating the planning of defense-based forts which account for around 3/4th of total existing forts in India.

Keywords Defense architecture · Fortification · Gwalior fort · Planning · Rajput

1 Introduction

The fortification tradition of India predates before kingdoms and empires were established although the oldest bastille may not have been built by people but hills, forests, and rivers were employed as lines of defense and refuges; those that were located on mountains or near water bodies are believed to be the best positioned to protect human communities (Deloche 2007).

The fortifications of Ancient Civilization of Indus Valley were built of mud, backed bricks, and stones and the town had two distinct divisions: citadel and lower town. The enclosure walls were pierced by elaborately designed gateways found in the Kalibangan, Harappa, and Surkotada (Deloche 2007). The rise of strong kingdoms

M. Tiwari (✉)
Aditya College of Architecture, Mumbai, India
e-mail: muskan.t@aditya-arch.edu.in

Vijayalaxmi J. (ed.), *Conservation of Built Heritage in India*, Advances in 21st Century Human Settlements, https://doi.org/10.1007/978-981-99-8335-3_5

made it imperative for developing stronger military and defensive capabilities. Also, the subsequent phase of urbanization saw a major expansion of the settlements in the Gangetic basin.

In the Kautilya Arthashastra, different categories of forts are mentioned like the *Dhanva Durg* or desert fort; the *Mahi* Durg or mud fort; the *Jala* Durg or water fort; the *Giri* Durg or hill/mountain fort; *Vana* Durg or forest fort; the *Nara* Durg or fort protected by loyal soldiers. Large human settlements are thought to be best protected by the mountain and river forts. Because of the defensive advantages that this natural barrier provided, the majority of the early forts were built on the stream path. This is the case, in the Kausambi on the Yamuna, Sisupalgarh on Mahanadi, and Nagarjunakonda on the Krishna River (Deloche 2007). The Shilpa sastra distinguishes different types of ramparts in the enclosure plans: square, round, rectangular, elliptical, triangular, and octagonal. This is the case in the Ujjain with irregular polygon, trapezium at Kausambi. Three main features of the defense system have evolved from the ancient technical treatises: the rampart ditch, the gates and the flanking towers.

Mauryan dynasty brought in the existence of the concept of Mud rampart using brick and stone parapets with military troops positioned along the rampart with moats filled in with lotuses and crocodiles. Sangam and Gupta periods witnessed the prosperity in natural, cultural, architectural, and political values as a pioneering testimony of the time.

Ibn Battuta, a Moroccan traveler describes the Gwalior fort located in the Gird region of Madhya Pradesh, India (Figs. 1 and 2), as an impenetrable fort which is an example of Giri Durg. This paper supports the defensive design strategy of Gwalior fort which has many palaces, temples, water tanks, entrance gates, and other monuments including jails, such distinctions made the fort historically germane.

The art and architecture of few Indian forts have been addressed by many authors (Deloche 2007, Burton 1960) but some have focused only on a specific fort type. (Gautam et al. 2012) Conducted scientific analysis of the built structures and (Perriere 2016) appreciated the art and architecture of the Gwalior Fort but neither of them has established the implications of its ideal defense system which is the Rajput's inherited citadel remains unknown and less investigated. Hence, this research paper will bridge the gap between them.

2 Study Area

2.1 Political History

The city and its fortress have been ruled by several historic northern Indian kingdoms. Suraj Sen (255-282AD) is considered the founder of the city named Gwalior. In the fifth century Huna tribe ruled the fort where they built a sun temple that does not exist today. They were defeated by the Gupta Empire, and later the city came under

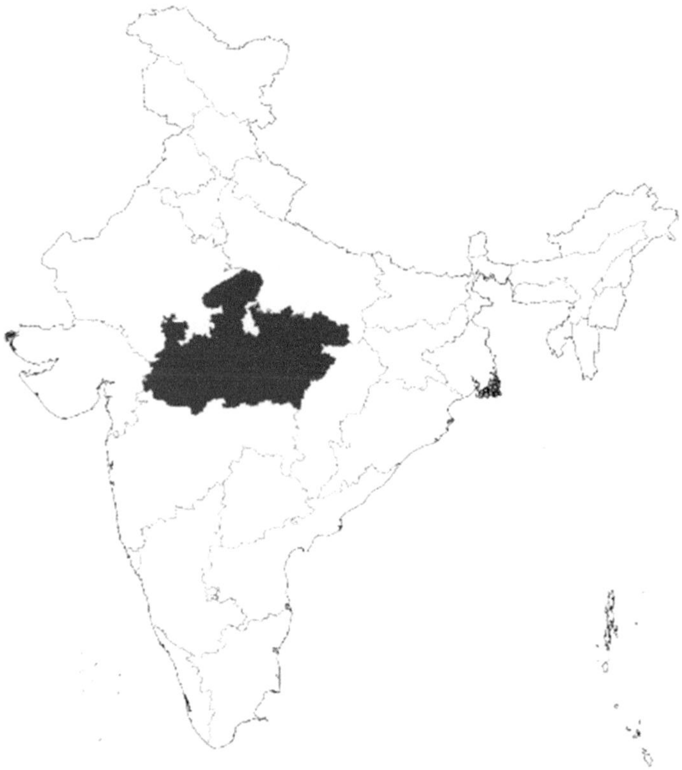

Fig. 1 Map of India with Gwalior city location. (*Source* https://www.britannica.com/)

Gurjara-Pratihara in the ninth century. Gurjara and Pratihara were originally pastoral and nomadic Rajasthani tribes who underwent major social transformation (Jadon 2019). Layers of fortifications are found from the Gurjara Pratihara time period in the fort. In the tenth century, the Kachchhapaghatas succeeded Pratiharas. In 1021AD Mohmud Gazni attacked and succeeded and was ruled by Mughals till 1210 when the Parihar Rajput regained the Gwalior fort. Later in 1232, the fort was succeeded by Iltutmish till 1375, when Raja Veer Singh was made the ruler of Gwalior, he was the founder of the rule of the Tomars in the city. They were the feudatories of the Tughlaq dynasty. In 1542, Sher Shah captured the fort after which the fort went under the Mughals. It was 1770, and the Maratha defeated Aurangzeb and set up the state of Gwalior (Jadon 2019).

From the ninth century itself, Gwalior was ruled by Rajputs and there were a lot of conflicts that happened between the Rajputs and the Mughals. After the attack of Mohd Gazni, the Parihar Rajputs focused more on strengthening their fort.

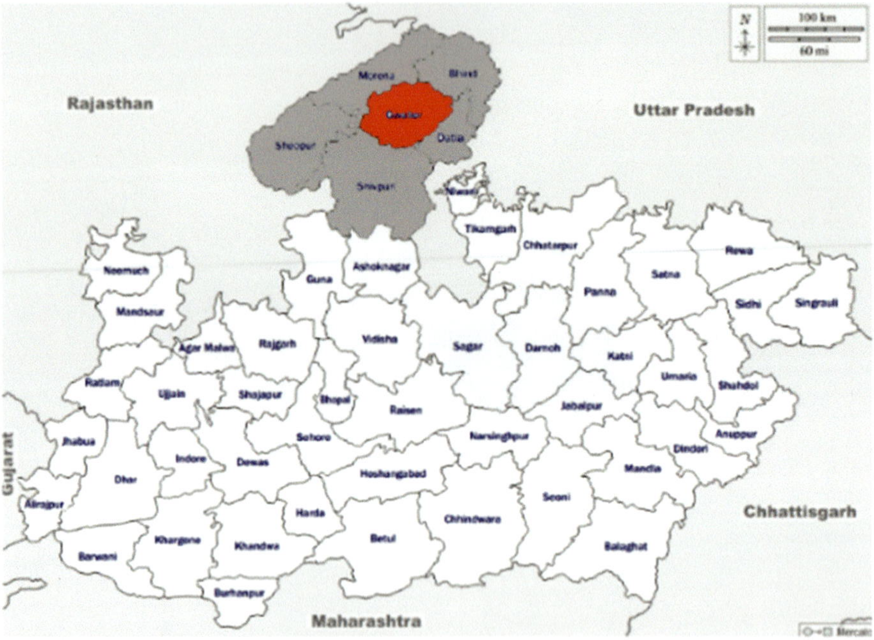

Fig. 2 Madhya Pradesh map of India depicting gird region and the Gwalior city. (*Source* Wikimedia.orghttps://www.wikimedia.org/)

2.2 Geography

The fort is situated on the top of a ridge of Vindhyan sandstone on a rocky mound called Gopachal Parvat. The ochre-colored sandstone rock formations of the Gwalior range are covered in basalt. It is situated on a rocky platform 104 m high at its highest point (Fig. 3), 2.4 km along its north–south axis, and 820 m from east to west (Fig. 4). The Swarnrekha River flows in proximity to the palace which is the main source of water for the moat built around the fort. There are several spring sources in the fort that act as the source of water and are able to offer water for a 15,000-strong garrison, the people required to protect the fort.

2.3 Architectural History

Inscriptions found in the vicinity of the fort have identified the fort construction to be from the early eighth century. There are two main access gates for the fort (Fig. 5)–one from the North East with curtain walls (Fig. 6) and the other from the South West. There are multiple entrances to the fort, out of which two main entrances exist today, one on the South West side, called Urvai Gate (Fig. 7) and the other on North

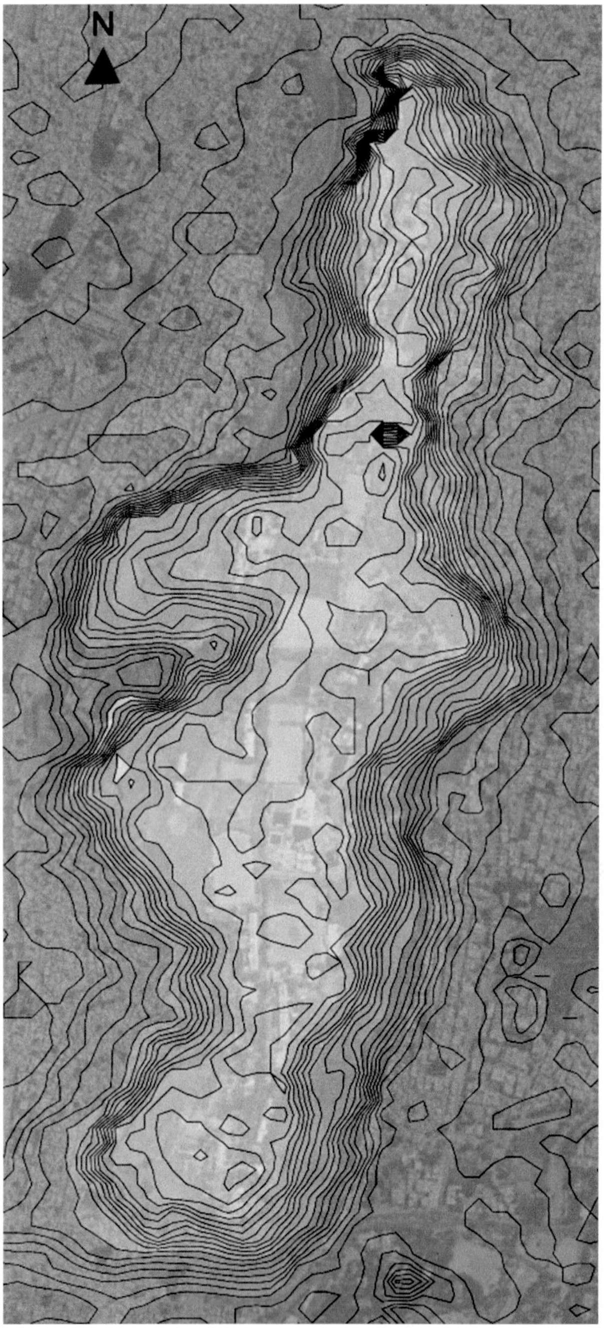

Fig. 3 Topographical Mapping of Gwalior Fort with the contour interval of 10 m prepared through GIS. (*Source* Author)

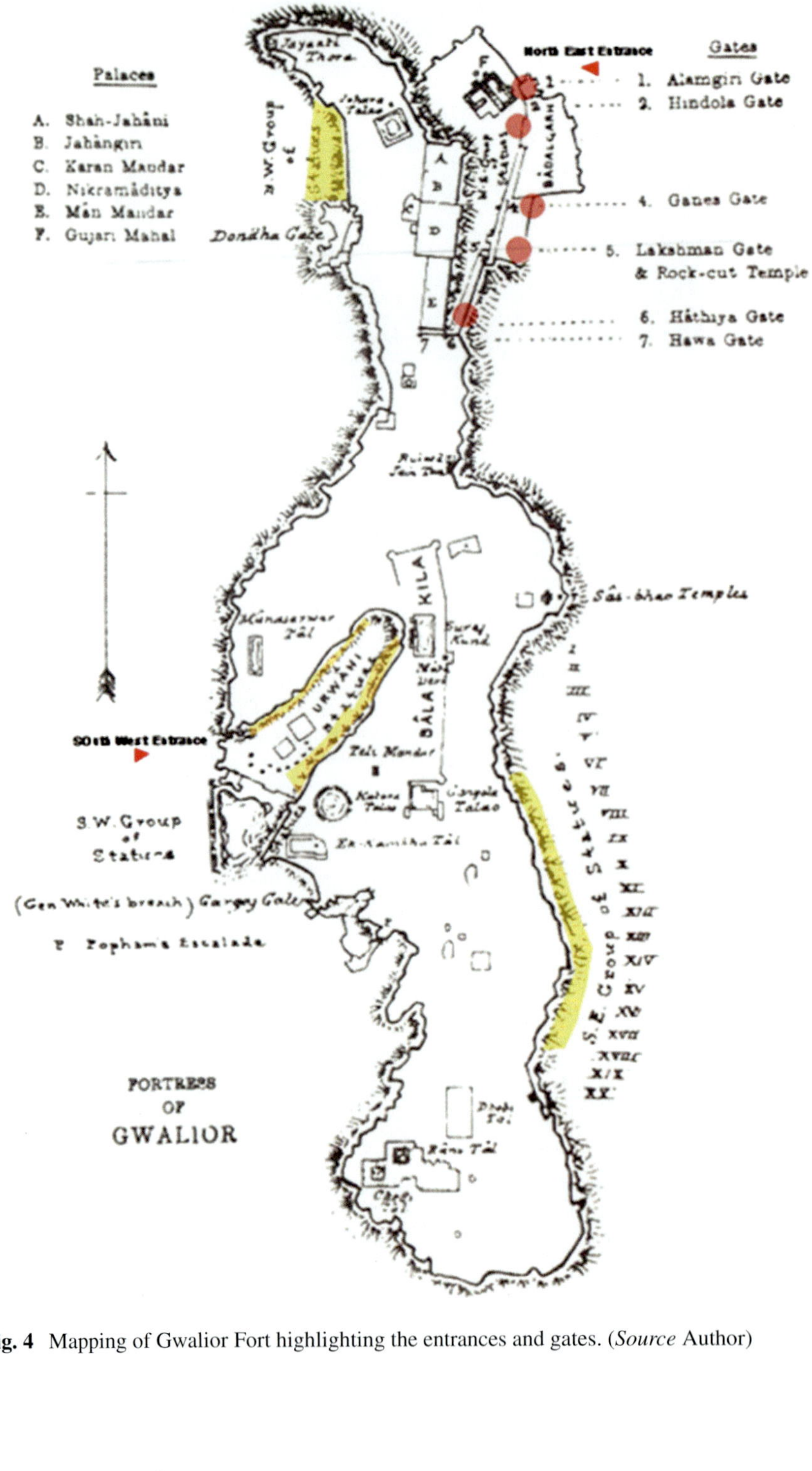

Fig. 4 Mapping of Gwalior Fort highlighting the entrances and gates. (*Source* Author)

East side called Hathi Pol or Hathi Darwaza (Fig. 8). The path to Hathi Pol was built by Man Singh and was named such due to the carving of an elephant placed there in the fifteenth century having multiple gates made by various kings over time, till the sixteenth century, with a separate platform built for firearms (Fig. 9). These gates include Alamgiri gate (Fig. 10) built by Muhtamad Khan, Governer of Gwalior in 1660, Hindola gate (Fig. 11), Bhairon and Ganesh Gate, both built by Dungar Singh of Tomar clan in the fifteenth century AD, and Lakshman Gate (Fig. 12) built in the ninth century by Laxman pal king of Kachhawaha ruler.

The majority of these gates not only had defense mechanisms but cultural associations associated with them which have various objects of historical and antiquarian interest. The path to Urvai gate has many huge Jain shrines carved on the stone of

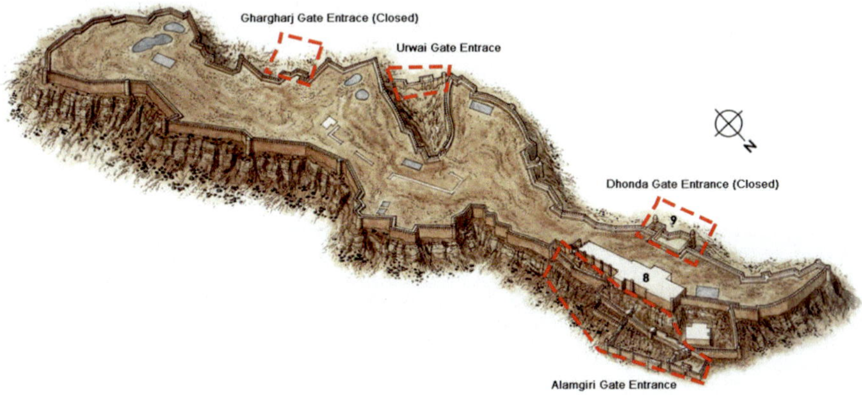

Fig. 5 Three-Dimensional Model of Gwalior Fort highlighting the entrances and gates. (*Source* Gwalior soulsteer.com)

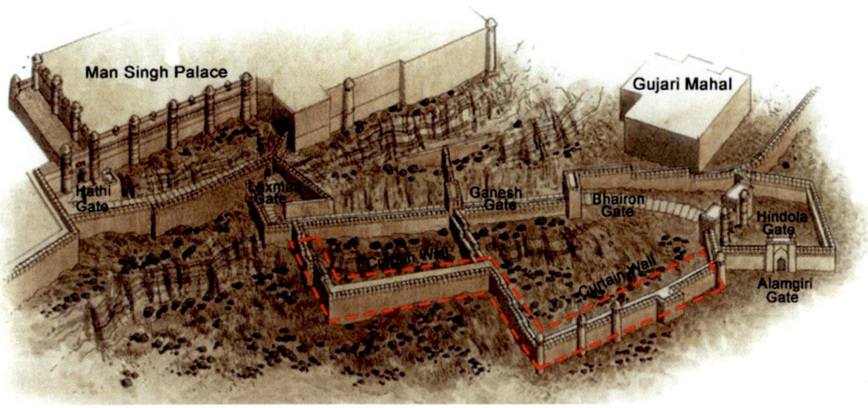

Fig. 6 North East Entrance detailed with curtain walls. (*Source* Gwalior soulsteer.com)

Fig. 7 Urwai Gate entrance through South West. (*Source* Author)

Fig. 8 Hathi Gate, small turrets projecting above the gateway. (*Source* Author)

Fig. 9 (Left) Platform for firearms outside Hathi gate. (*Source* Author)

Fig. 10 (Right) Alamgiri Gate, entrance through Northeast. (*Source* Author)

Fig. 11 (Left) Hindola Gate, Towers projecting outwards. (*Source* Author)

Fig. 12 (Right) Laxman Gate, entrance through Northeast. (*Source* Author)

the hill which dates to the 6th to fifteenth century AD (Costello 2020). Also, outside Lakshman gate, there is Chaturbhuj temple.

There are various structures and sculptures like the sculptures of Jain Tirthankaras, Teli ka mandir built in the eighth century by Pratihara king known for its Vallabhi style, Sas Bahu temple (1092-93AD) built by the Kachchhapaghata dynasty, Man Singh Palace built by Man Singh in 1508AD, Karan Mahal, Vikram Mahal, Gujri Mahal, Jahangir, and Shah Jahan Mahal, Bhim Singh Rana water tank and Cenotaph built in 1740, Chaturbhuj temple build by king Bhoja in 875AD build on a monolithic rock-cut temple known for its earlier existence of zero (0) symbol found in its inscription and Gurudwara Data Bandi Chhor.

3 Methodology

To find an appropriate methodology for the present research paper, several books (Deloche 2007), research paper (Burton 1960) and PhD thesis on Defense systems and fortification were referred to with Archival records supported by field studies and primary information from government agencies and historians used in this study. This research specifically focuses on the:

- Political history of the region with reference to the Tomars and Mughals.
- Defense weapons used during this period by the rulers.
- Defensive features inculcated in the Gwalior fort to prevent the fort from war.
- Impact of defensive technique on the planning of Gwalior fort.

The paper considers both architectural and defense attributes in order to identify their impact on the planning of the fort.

4 Defense Architecture of the Fort

In the fort from the eleventh century till the seventeenth century, there were constant conflicts between the Muslim and Rajput dynasties. It was the time when gunpowder and firearms came into use, hence defense systems were strengthened in the fort. The most significant defense feature of the fort is the series of six gates from the east side of the fort. Different rulers were the followers of different religions; hence, a wide range of cultural diversity can be seen on the fort. There are numerous places of worship that are associated with the belief of praying the god before the battle by the kings and day-to-day by the queens and the royals. Under the Mughal rule, the fort witnessed the construction of mosques like Shahi Jama Masjid and under Hindu, there were temples like Sas Bahu and Teli ka Mandir. The rulers focused on building forts with:

Fig. 13 (Left) Wall walks were added along the top of the ramparts. (*Source* Author)

- Wall walks were added along the top of the ramparts (Fig. 13).
- Thick stone ramparts followed by Merlons at the parapet (Fig. 14) for the protection of the soldiers with loopholes in between them (Fig. 15).
- Gates which were considered the weak spot in defense developed as—rows of sharp, stout iron spikes.
- Small turrets projecting from the parapet of walls.
- Cavalier, or gun platforms were raised on towers.
- Bastions all along the fort wall of length 12 km were constructed.
- Lower enclosure was constructed all around the main one, with parapets pierced by loopholes (Fig. 16).
- Wet moats (Fig. 17) were constructed.

With the progress of defense weapons like artillery to the introduction of gunpowder, lowering the curtain walls and the towers was never considered by the military engineers to reduce the surface exposure which may have helped to minimize their impact, if we compare it with Western countries where they shifted to trenches and low extended parapets.

Fig. 14 (Right) View of Dhonda Gate. (*Source* Author)

Fig. 15 (Left) Loopholes after Firearms. (*Source* Author)

4.1 Defense History

From the eleventh century till the seventeenth century, there were constant conflicts between Muslim and Rajput dynasties in central India. It was the time when the

Fig. 16 (Right) Curtain wall on the North East side of the hill. (*Source* Author)

Fig. 17 Traces of wet moat and Urwai gate entrance. (*Source* Author)

Fig. 18 (Left) Bastions and Merlons, Gwalior Fort. (*Source* Author)

defense systems were given more importance than the palace and religious architecture. The Rajput rulers focused on building forts with massive stone or earthen ramparts followed by wet and dry moats, bastions (Fig. 18), and merlons (Figs. 18 and 19) as parapets for the protection of the soldiers with loopholes in between them (Fig. 21), battered walls acting as a retaining wall against canon ball attacks, curtain walls (Fig. 20) and gates with rows of sharp, stout iron spikes to dissuade an attacking army from using elephants to break down the gates.

4.2 Defense Systems

The third century till the fourteenth century AD was the time when gunpowder and firearms were not in existence. As a result, the fortifications at the time were made up of towering, massive, and compact earthen walls covered in stone facing made of enormous blocks laid without mortar, strengthened at regular intervals by solid quadrangular towers, crowned by battlements, provided with strong entry points with open courtyards between, and bounded by broad and deep ditches. (Deloche 2007).

With the arrival of the Muslims, a robust style of military architecture developed in the Deccan, and the use of firearms in the sixteenth century led to further advancements in the art of defense. As a result, most strongholds built in the Deccan kingdoms now feature more effective defense works, such as circular or semicircular towers, enormous gates guarded by barbicans, and machicolations in the vault of a

Fig. 19 (Right) Merlons, Gwalior Fort. (*Source* Author)

Fig. 20 (Left) Bastion elevation, Gwalior fort. (*Source* Author)

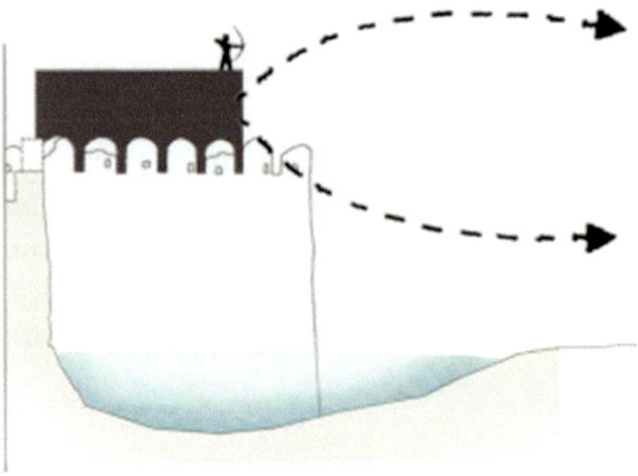

Fig. 21 (Right) Bastion Plan, Gwalior fort. (*Source* Author)

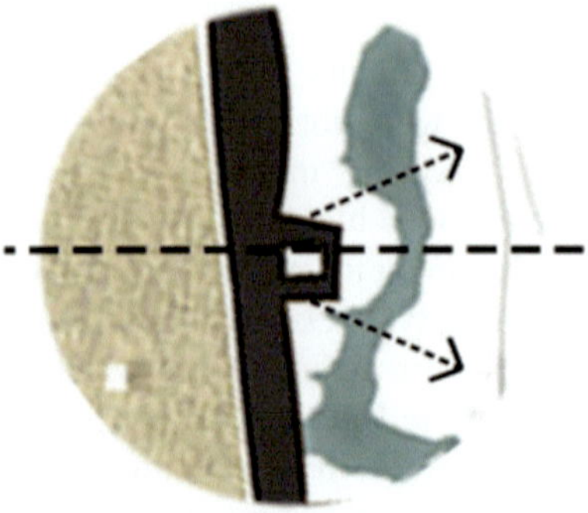

Fig. 22 Merlons, Gwalior fort. (*Source* Author)

gate, through which attackers could drop stones or burning objects. There are also spherical stone balls known as cannonballs found in certain parts of the fort.

During the fifteenth century when the Mughals captured Gwalior fort, they did not destroy the old strongholds of stone revetments but had to adapt the existing works of the Hindu engineers with modifications in it. Every time a new development in weapons occurred, they adjusted the higher portions of the walls and left the walls of the curtain walls and towers untouched. As a result, the fort's historic battlements were replaced with strong, over three-meter-tall merlons on all its walls (Fig. 22), and the quadrangular towers (Fig. 23) with cavaliers topped on it, within which artillery pieces were installed.

4.3 Defense Management

Troops were recruited, equipped, and trained by the state (*Maula*). There were several groups and forest tribes (*atavika*) that were recognized for their military skills and valued as such like Bandits of Chambal, Thugs of Guna & Shivpuri, Gond of Malwa, etc. These people made their living as soldiers (*ayudhjivi*). Communities that supplied soldiers were known as ayudhiya. Also prevalent were corporate guilds of warriors (*shreni*) and mercenaries (*bhrita*), who were hired as needed. The army was made up of infantry, cavalry, chariots, and elephants (*chaturanga*) (Deloche 2007). They

Fig. 23 Gateways topped by quadrangular towers. (*Source* Author)

were all placed in formation (*vyuha*) on the field of battle as determined by the commanders based on variables including the geography and the make-up of their own and the enemy's troops.

4.4 Defense Weapons

Initially, arms included swords, bows and arrows, double-handed broadswords, rectangular, oval, or bell-shaped shields, javelins, spears, lances, pikes, axes, clubs, and maces (Fig. 24). Bows were the main weapon for the chariot, infantry, elephant warriors and the commanders. Before the appearance of the gun in the thirteenth century, weapons such as fire arrows, bombs, and the fire lance were in use since the tenth century. According to the presentation of Iqtidar Alam Khan, gunpowder appears to have arrived in India during the second half of the thirteenth century from China, a type of rocket was subsequently incorporated in India and north of the Deccan in the second half of the fourteenth century, and the use of gunpowder artillery increased during the middle of the fifteenth century. In 1498, after the advent of the Portuguese and Babur's invasion, guns became an important part, and were widely used.

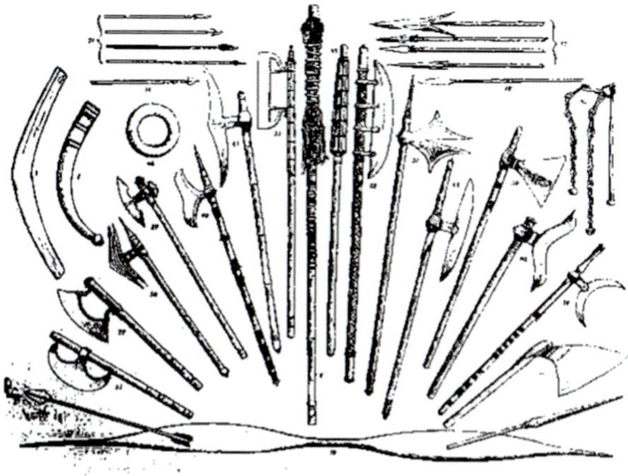

Fig. 24 (Left) Weapons used in war before the introduction of firearms, Gwalior. (*Source* Unknown)

Fig. 25 (Right) Cannons in Man Singh palace, Gwalior. (*Source* Author)

Besides military use, gunpowder was also being employed as explosives for tunneling and mining or as display fireworks. These forts have two different kinds of cannons: the first is a cast-bronze cannon, and the second is a wrought-iron cannon made of iron bars and rings. (Fig. 25).

4.5 *Defensive Features in the Fort*

They were necessary for the security of the entire kingdom as well as the people who lived nearby since they protected the monarch and his soldiers from foes and prevented attackers from moving deeper inland. Forts had to be conquered as enemy capitals were frequently fortified and no invader could declare victory until he had taken control of these vital fortifications. Moreover, forts were used as sites for administrative divisions. The gate of the fort has steel spikes to protect it from elephants and no direct entrance was given through the straight road but a curvilinear road leading to the gate to protect the door from the forces of wooden logs. Above the gate, there is generally a room that houses guards to watch over the entrance or shoot from the gaps/holes in them. Also, the forts were built as such to reduce the speed of the horses and to trap the enemies in the courtyard lying between the two gateways.

5 Discussion and Conclusion

The Gwalior Fort is a piece of architectural marvel in the state of Madhya Pradesh, India. It has a rich and storied history that spans over a thousand years. The Gwalior Fort is believed to have been built in the eighth century by the rulers of the Kachchhapaghata dynasty. The fort's strategic location atop a steep hill provided a natural defense against invasions. The Tomara Dynasty, the Rajput Dynasties, the Mughal Empire, the Scindia Dynasty and the British rule, all had control over the fort until India achieved its independence in 1947. Since then the fort has been maintained by the Archaeological Survey of India.

During the Indian Rebellion of 1857, the Gwalior Fort became a significant center of rebellion against British rule. The Rani of Jhansi, Lakshmibai, played a crucial role in the uprising. The fort is an amalgamation of Hindu, Jain and Islamic architecture with several notable structures resembling varied architectural styles. Yet, its truly significant feature is its defense architecture, which needs to be studied further in depth to align with the times of the inception of the defense systems. The fort has witnessed numerous historical events and remains a symbol of Gwalior's rich heritage.

The fort was identified as significant by all the rulers for centuries. Hence, various layers of fortification are visible. The various defense structures that existed inside the fort were built with specific materials and architectural designs that demonstrate the fort's strategic thinking, military prowess, and judicious practicality. It was constructed as a luxury palace protected by religious ideals as well as a military observation point for protection. From the time it was built in the eighth century till the eighteenth century, it was a fully operational fortress, which attested to its durable and strong nature and, hence, the impenetrability of the fortifications. The design of the fort at Gwalior is evidence of the rich history and achievements of the Rajput and Mughal dynasties.

References

Burton J (1960) A study of fortification in the Indian subcontinent from the thirteenth to the eighteenth-century A. D. Bull School Orient African Stud 23(8):508–522. https://www.jstor.org/stable/610034

Costello C (2020) Inscribing the social and sacred landscape: jain monuments and their memorial traces in medieval Gwalior. South Asian Hist Cult 11(1):60–77. https://doi.org/10.1080/19472498.2020.1719755

Deloche J (2007) Fortifications in Ancient India, studies on fortification in India, pp 15–25. https://books.openedition.org/ifp/4014

Gautam PK, Singh D, Bhadauria S (2012) Effects of growth of angiosperms on the monuments—investigation based on historical monuments of Gwalior Fort. Nat Environ Pollut Technol Int Quart Sci J 11(2):307–309. https://neptjournal.com/upload-images/NL-38-23-(23)B-1843.pdf

Jadon SS (2019) Understanding with the chronological events: historical layering of the old Gwalior. JETIR 6(2):672–682. https://www.jetir.org/papers/JETIR1902088.pdf

Perriere EB (2016) Gwalior fort art and architecture. Indian Archaeol Hist Art 136–139. https://www.academia.edu/42806157/Gwalior_fort_art_and_architecture

https://commons.wikimedia.org/wiki/File:India_Madhya_Pradesh_location_map.svg. Accessed 23 May 2023

https://gwalior.soulsteer.com/2021/06/24/birds-eye-view-of-excellent-fortification-of-gwalior-fort-with-its-map/. 23 May 2023

https://www.bibliotecapleyades.net/vimanas/esp_vimanas_11c.htm. 23 May 2023

https://www.britannica.com/place/Madhya-Pradesh. Accessed 23 May 2023

Holistic Study to Comprehend Influence of Orissa and West Bengal on Temple Styles in Nagpur

Yogini Moreshwar Damle

Abstract The temple architecture of the Nagpur region in Maharashtra, India followed a very peculiar style as compared to the other regions of Maharashtra. The architectural character of the temples of the Nagpur region was greatly influenced by the Orissa and West Bengal styles of temple architecture. This is because temples of Nagpur during the eighteenth century in India have been supported by the Craftsman of the neighbouring states of Orissa and Bengal. Nagpur was predominantly ruled by the royal lineage of Bhonsale of the Kolhapur region. Their feudatories also called Bhonsales were from the Kolhapur region of Maharashtra. The research paper aims to assess and analyse the influence of Orissa and West Bengal temple architecture styles on the Bhonsale-ruled temple architecture of Nagpur. A comparative study of the planning features and architectural characteristics of the temples of Nagpur region and other parts of Maharashtra is carried out to assess the influence of Orissa and West Bengal temple styles. To understand how the Nagpur temple style is similar and collect data on Bhonsales of Nagpur and their expeditions, meetings with the officials from Orissa and West Bengal. This study is important to assess the contribution of Craftsmen from Orissa and West Bengal which has resulted in a unique style of temple architecture in Nagpur since such studies are limited.

1 Introduction

Nagpur is located in the northern part of Maharashtra which is called a central province in India. During the eighteenth century, it was ruled by Hingnikar Bhonsles of Nagpur. The Bhonsale family is the royal or Kshatriya clans of the Marathas (Apte).

Marathas were ruling some parts of the Karnataka region in the eighteenth century as well as parts of Rajasthan and Gujarat so there are evident examples of Karnata

Y. M. Damle (✉)
SPAV, Vijayawada, India
e-mail: damle26yogini66@gmail.com

Vijayalaxmi J. (ed.), *Conservation of Built Heritage in India*, Advances in 21st Century Human Settlements, https://doi.org/10.1007/978-981-99-8335-3_6

dravida and Nagara shikhara in different parts of Maharashtra, other than Nagpur region, due to the involvement of craftsmen from those regions.

In various regions of Maharashtra, we can find Karnata Dravida shikhara and a few Nagara language Shikhara forms with typical timber Mandapas. Whereas in Nagpur even though the language which was followed by the other parts of Maharashtra can be seen, the style of details is very different and is peculiar to the Vidarbha region of the same time period.

2 Study Area

2.1 Context of the Area and History

Currently, Maharashtra is divided into 6 divisions namely Konkan division, Pune division, Nashik division, Aurangabad division, Amravati division, and Vidarbha division (Fig. 1). We can find a completely different style of temple architecture in the Vidarbha division than the other five divisions in Maharashtra.

The most experimentation of this style of architecture followed in Vidarbha division can be seen in Nagpur, where Hinganikar Bhonsales ruled. Bhonsales ruled Nagpur from 1743 to 1853. Temple architecture flourished in Nagpur under the rule of Hingnikar Bhonsales as they were great patronages for Art and Architecture (Gupte and Oak 2016).

The Bhonsale family that belonged to Chhatrapati Shivaji, the founder of the Maratha State, was from Verul, close to Daulatabad, formerly known as Devgiri. The Bhosles of Nagpur are referred to as Hinganikars because one of their ancestors, who was likely a contemporaneous of Maloji, Chhatrapati Shivaji's grandfather, restored the village of Berardi near Hingani in the current district of Poona. Sahaji Bhosle, the father of Shivaji, was a contemporary of the two brothers Mudhoji and Rupaji of Hingani-Beradi. The Hingnikar Bhosle family, like the Chatrapati Bhosle family, believes that it is descended from the Sisodia Rajputs of Udaipur. It is also possible that during the lengthy ascension of the Muslims, certain Ksatriya clans of the Rajputs descended to the Maratha territory from the north. Yet, it is a historical fact that Ksatriya families without ties to the north's Rajputs existed in the Maratha nation, including the Rastrakutas, Chalukyas, and Yadavas.

Hingnikar Bhonsale was Sena Saheb Subha for the court of Marathas and the Karnatak expedition of 1744 gained Raghuji Bhonsale a par status with that of Peshwas of Pune.

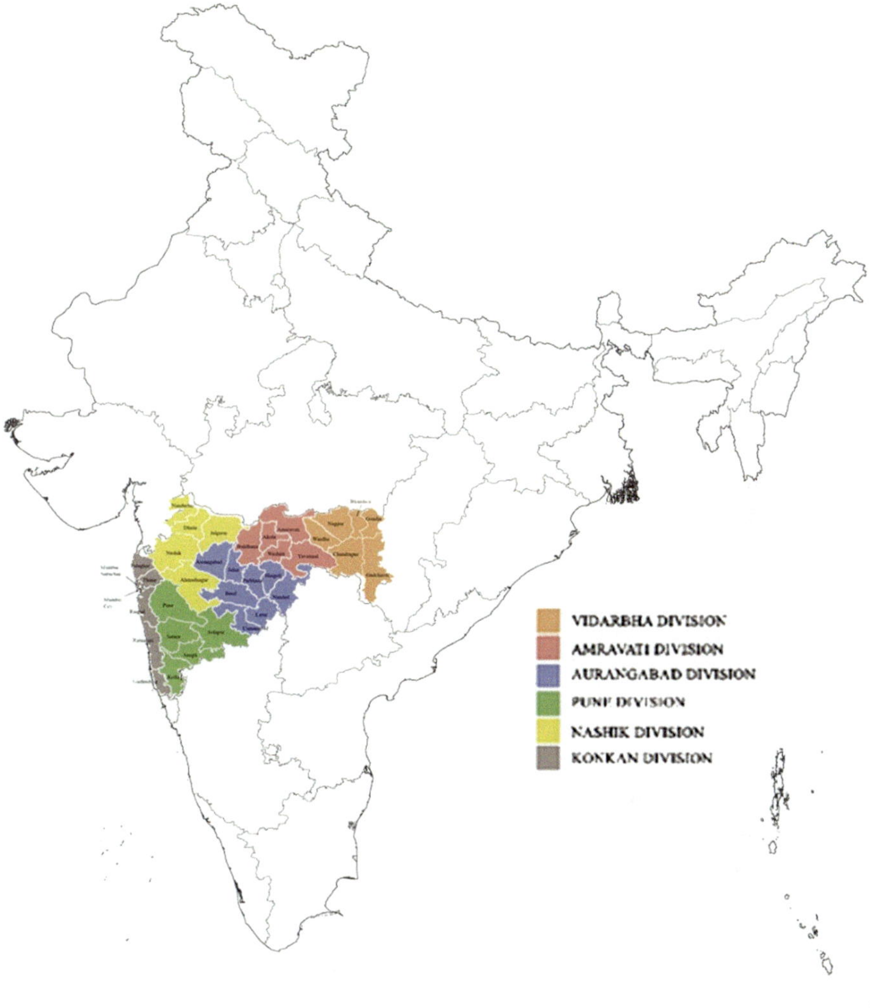

Fig. 1 Map showing the current political division of Maharashtra and the location of Nagpur

2.2 Extent of Rule of Marathas in the Eighteenth Century and Its History

During the rule of Yadavas of Devagiri in the eleventh century who ruled the Marathi speaking linguistic boundary, a new temple architecture style came up called as Hemadpanti named after the inventor of the style and minister of the Yadavas called as Hemadri. This temple style flourished till the end of Yadavas at the hands of Allaudin Khilji and Malik Kafur in 1303.

The following years came with the rule of different Islamic rulers and years of anarchy in the region until Chatrapati Shivaji Maharaj founded the Swarajya of Marathas. Even with the change in rule, it was the time when Maratha army focussed on establishing a proper rule and strengthening defences and their territory (Fig. 2). The flourishment of art and architecture in Maratha dynasty began with the rule of Chhatrapati Shahu Maharaj as it was the time of political stability and Marathas were one of the powerful dynasties at that time (Maharashtra State Gazetteers: Bhandara 1979).

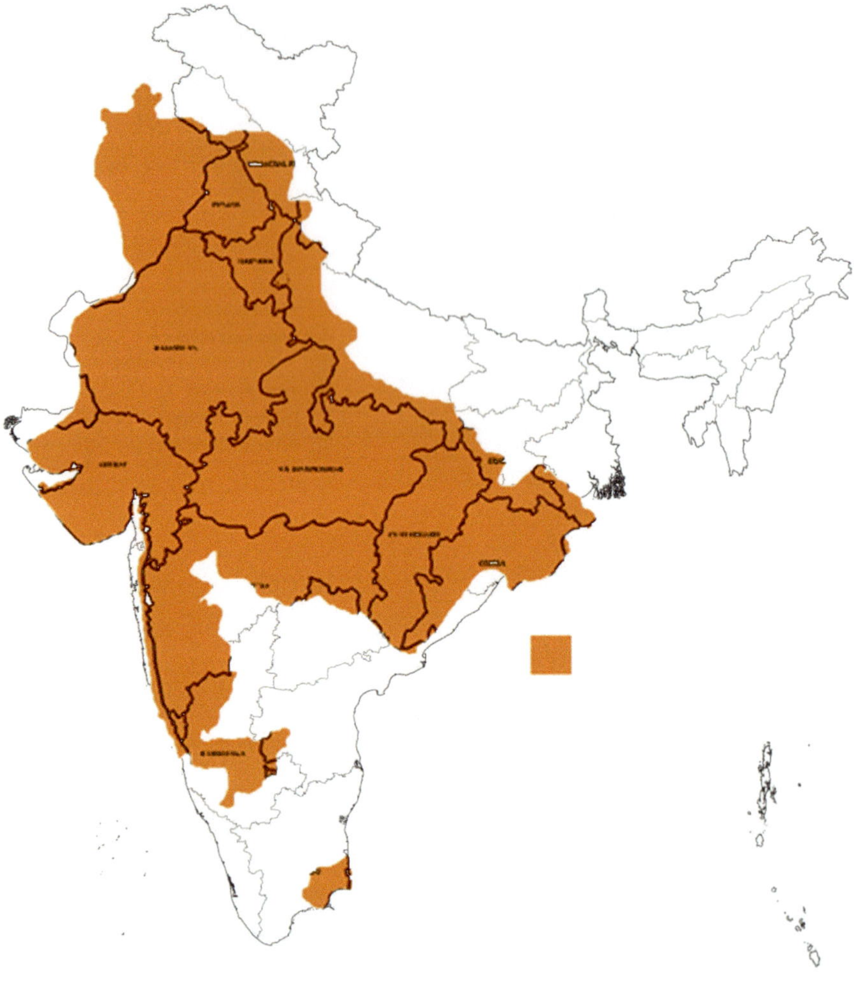

Fig. 2 Map showing the extent of Maratha rule in the eighteenth century. (*Source* Author)

Chhatrapati Shahu Maharaj grandson of Chhatrapati Shivaji Maharaj ruled most of the Indian peninsula in the eighteenth century with the help of his able ministers and generals. With the help of Bajirao Peshwa I and other ministers, he was able to get Madhya Pradesh, Rajasthan, Gujarat, Karnataka, Haryana, Punjab, Himachal Pradesh, Chhattisgarh in current India and parts of Tamilnadu under Maratha control. Some parts of current Afghanistan were also under the rule of Marathas. Raghuji Bhonsale I gained regions of Orissa and West Bengal for the Chatrapati. Thus the Marathas ruled most of the current India in the eighteenth century. We can find influences of these regions in the temples of Maharashtra built during the eighteenth century.

2.3 Extent of Hinganikar Bhonsales Domain in Maratha Confederacy in the Eighteenth Century

After disputes over the succession following the death of Chand Sultan, the Gond ruler of Nagpur, in 1739, Wali Shah, an illegitimate son of Bakht Buland Shah, took the throne. In order to protect the interests of her sons Akbar Shah and Burhan Shah, Chand Sultan's widow requested the support of Maratha leader Raghuji Bhonsle of Berar. The legitimate heirs to the throne were installed when Wali Shah was executed. Raghoji I Bhonsle was dispatched back to Berar with a substantial reward for his assistance. Seeing the splendour of his award, the Maratha general concluded that Nagpur must be a prosperous and abundant nation.

Nonetheless, the brothers' disagreements persisted, and Burhan Shah, the older, once more asked Raghuji Bhonsla for assistance. Akbar Shah was forcibly exiled before being poisoned in Hyderabad. While it was within his grasp, Ragoji Bhonsle did not possess the will to flee such a rich and abounding nation this time. He proclaimed himself the Gond king's "protector." As a result, Burhan Shah was essentially made a state pensionary in 1743, with the Maratha monarch holding actual power. The history of the Gond kingdom of Deogarh after this incident remains unrecorded. With the removal of the Gonds from the kingdom of Nagpur, a succession of Maratha kings, beginning with Raghoji Bhonsle, ascended to power.

The Nagpur-based Bhonsles were close family members of Chhatrapati Shahu, who gave them their rise to wealth and power. During the Alivardi Khan era, Raghoji I Bhonsle conquered Bengal and Bihar while seizing control of Orissa from the Nawab. They often opposed the Peshwa and asserted their own independence because they effectively held the Gond monarch of Nagpur under their power (Fig. 3).

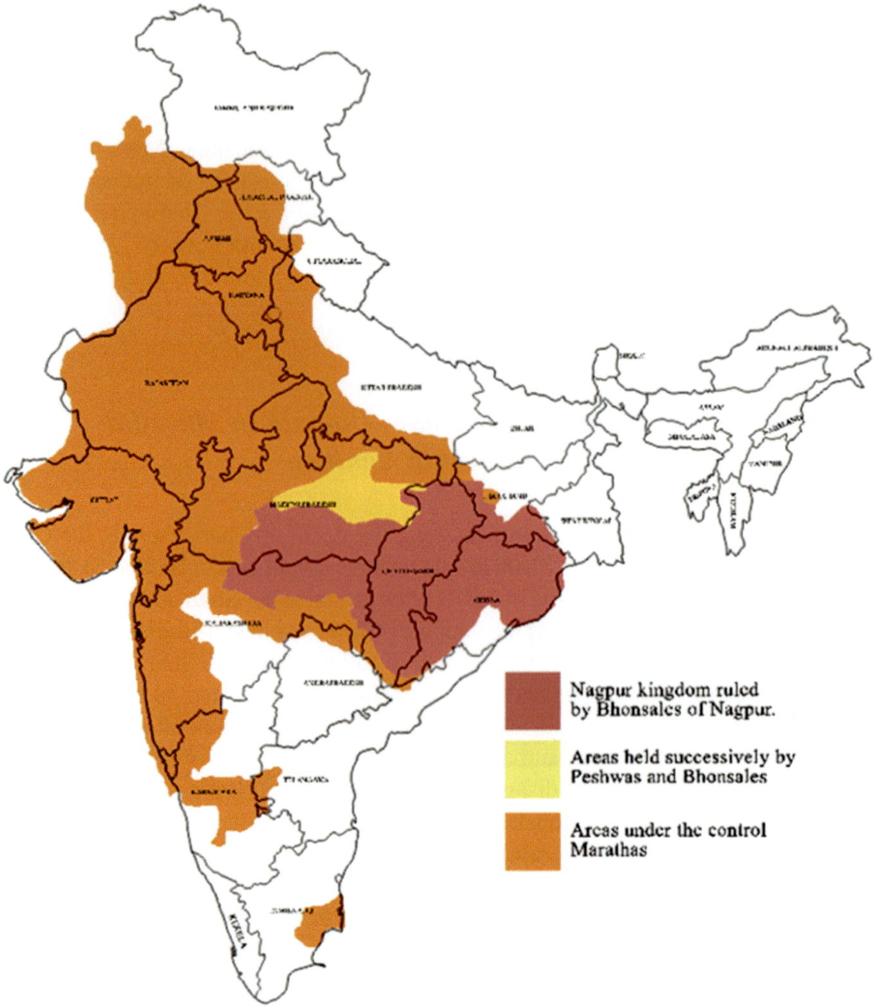

Fig. 3 Map showing areas held by Hingnikar Bhonsales and Peshwas

2.4 Raghuji's Raids on Bengal and Orissa

Bengal's political situation was unstable by the year 1740. The navab, who resided at Murshidabad, then ruled over Bengal, Bihar, and Orissa. Murshid Quli khan, a capable governor, passed away in 1727. An ambitious Turk in his service by the name of Alivardi Khan killed his son-in-law Sarfaraz Khan, who was the Navab, in 1740. The supporters of the deceased Navab despised Alivardi's coup. A capable Persian from Siraz by the name of Mir Habib led the powerful faction at Alivardi's court. Mir Habib had come from nothing to become the deputy navab-ship of Orissa.

If he launched an invasion, he had made proposals to Raghuji in Bengali territory. Raghuji, who had been waiting to expand his area of influence to the east of Nagpur, found this to be a highly alluring offer. Instead, he saw Bengal and the area between Nagpur as his particular area of expertise. His claim was enhanced by his outstanding achievements in Karnataka, and Chatrapati Sahu wholeheartedly supported it by giving him a sanad.

Mir Habib had gone to Nagpur to urge Bhaskar Ram to attack Bengal when Raghuji was in Karnataka. But, Bhaskar Ram waited till his master arrived back at his house from the far-off Karnataka. On his return from Karnataka, Raghuji made careful preparations and dispatched a 10,000-person force under the capable leadership of Bhaskar Ram. Bhaskar Ram embarked on the voyage on the auspicious Dasara day of 1741.

After many ensuing fights between the Nawab, Hingnik Bhonsales, and Mir Habib the expedition of Bengal and Orissa of Raghuji came to an unusual pause because, The Peshva had departed Poona in 1741 with the intention of suppressing Raghuji's operations in Bengal. With the aid of Malharav Holkar, he strengthened his position in Malva and conquered Gadha and Mandla while robbing Sivani and Chapar. Alivardi was extremely alarmed upon learning of the Pesva's operations because he anticipated a united attack by the Pesva and Raghuji. In exchange for being given the power to govern Malva, Bundelkhand, and Allahabad, the Pesva vowed to aid the emperor and Alivardi Khan against Raghuji. This suggestion was immediately accepted by the Emperor, who then dispatched the Pesva to remove Alivardi.

Balaji and Raghuji were summoned to Satara by Chatrapati Sahu, who was aware of their long-standing animosity towards one another. He facilitated a reconciliation that was respected by all parties. Neglecting the breach would undoubtedly have hurt the interests of the Maratha authority in India. On August 31, 1743, Raghuji and Balaji signed a contract in Satara in front of the Chhatrapati. By doing this, all of the area east of this line, including Ajmer, Agra, Prayag, and Malva, was given to Raghuji, and the land west of this line, including Katak, Bengal, and Lucknow, was given to Balaji Pesva. Nobody was to enter another person's space. Thus Raghuji continued his activities in Bengal and Orissa after returning from Satara and it ended with Alivardi signing a treaty with Raghuji Bhonsale.

For more than a couple of years, Janoji was in Orissa and Raghuji was busy with the political affairs at Satara and Nagpur. The treaty was signed in May, 1751:

(1) Mir Habib was to be confirmed in the Government of Orissa as the deputy Subhedar of Mursidabad.
(2) The Navab was to pay annually 12 lacs of rupees to the Bhosles of Nagpur for the cauthai of Bengal and Bihar.
(3) When these amounts were regularly paid, the Bhosles were not to harass the two provinces.
(4) The district of Katak i.e., the territory up to the river Suvarnarekha was to be considered as the possession of the Bhosles.

Thus, Hinganikar Bhonsales might have had a major influence on Bengal and Orissa over their architectural style because of their extensive expeditions and stay in this region.

Nagpur's prosperity can largely be attributed to Raghuji. He brought several Maratha and Brahmin families from western Maharastra with him, and they gave new order and vigour to Nagpur and Berar's government. During Raghuji, Nagpur cultivation improved. Families of cultivators, or Kunbi, began to settle in the area ruled by Raghuji. Raghuji Bhosle deserves credit for paying the Kostis or weavers. Rama was a deity to Raghuji. He restored the significance of this site of legendary repute by installing the Rama idol there. Many more temples and holy sites received land concessions from him.

3 Temple Typology of Nagpur

3.1 Introduction to Temple Typology of Nagpur

From (Figs. 4, 5 and 6)of Murlidhar Temple, of Chitnavis Wada in Nagpur, we can see that the Mandapa part of the temple was constructed in timber and the antarala and shikhara part of the temple was constructed mainly in red and yellow sandstone. We can see the use of bricks in some temples for the construction of the roofs of antarala and mandapas. In some cases, we can see the use of lime plaster to plaster the walls.

Fig. 4 Shikhara of Murlidhar temple of Shekhari mode. (*Source* Author)

Fig. 5 Antarala of Murlidhar temple. (*Source* Author)

Fig. 6 Mandapa of Murlidhar temple. (*Source* Author)

The plan form is divided into three parts that is Sabha-mandapa, then antarala and gabhara. The two sides of Antarala are sub-shrines dedicated to either demi gods or mainly Ganapati and Shiva. In Fig. 7 of Murlidhar Mandir we can see that following the same flow of spaces the sub-shrines are dedicated to lord Ganapati and lord Shiva. From (Figs. 8, 9, and 10) various modes of shikharas can be seen in Nagpur, such as Latina, Shekhari, Bhumija and a few Karnata dravida and also we can see elements of Bengali style curved roofs in few temples. Murlidhar Mandir has Shekhari mode of Shikhara of Nagara language (Fig. 11).

The mandapa portion of the temple is constructed in timber and is usually flat in the centre and sloping towards an outer row of columns (Fig. 12) but the mandapa roof of Murlidhar temple is constructed as a pitched roof (Fig. 13). The Mandapa roof of Murlidhar temple is an exception as it is speculated that it was renovated at the

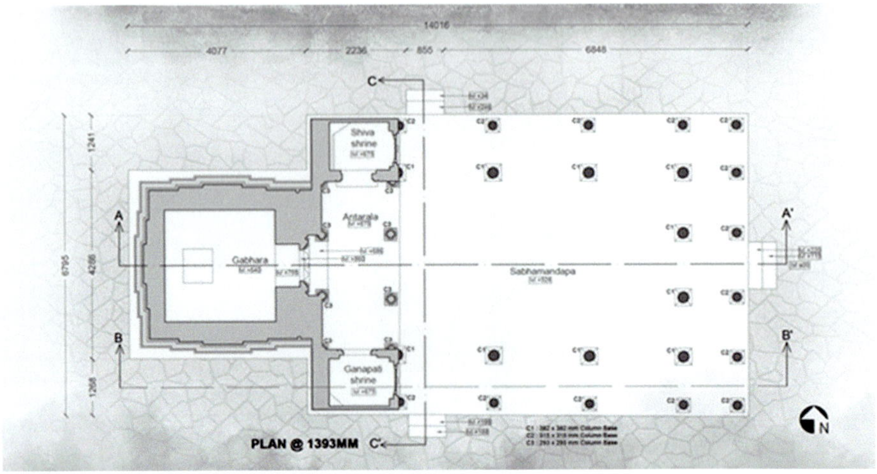

Fig. 7 Plan of Murlidhar Mandir. (*Source* Author)

Fig. 8 Shikhara of Murlidhar temple Sonegaon of Latina mode. (*Source* Author)

start of the nineteenth century, otherwise in all the temples constructed by Bhonsales we can see a flat roof over the mandapa or a little bit slope at the outer column. A comparison of the temples from other parts of Maharashtra and Nagpur is shown in Table 1

The columns in Mandapa are usually of timber and columns in antarala are of sandstone (Fig. 14) but there are a few exceptions where mandapa columns are made of sandstone along with the antarala columns (Fig. 15).

Fig. 9 Shikhara of Rukmini temple of Bhumija mode. (*Source* Author)

Fig. 10 Shikhara of Ghui temple with hint of Karnata Dravida. (*Source* Author)

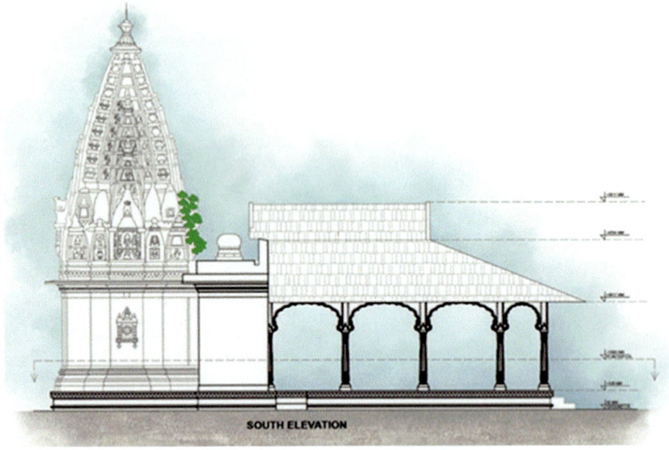

Fig. 11 South Elevation of Murlidhar Mandi. (*Source* Author)

Fig. 12 Mandapa roof of Ganesh Mandir Shukravari. (*Source* Author)

Inner row of the column is longer in length and outer row is smaller to make the roof sloping towards outer row (Fig. 16).

Comparative analysis in Vidarbha region and other areas of Maharashtra (how temples in Vidarbha are unique to the region).

From the above comparative analysis, we understand that there is very slight difference between the temples from the other regions of Maharashtra and temples of Nagpur. The differences are stylistic in nature and are very minute to be noticed. The material change is also noticed. Thus, temples constructed in Nagpur in the

Fig. 13 Mandapa roof of Ghui temple. (*Source* Author)

Table 1 Comparative features of temples from other regions of Maharashtra and Nagpur

Temples from other region of Maharashtra	Temples from Nagpur
 Mandapas in other regions of Maharashtra have timber columns, the planning of these mandapas may vary, and we can find two storeyed mandapas too	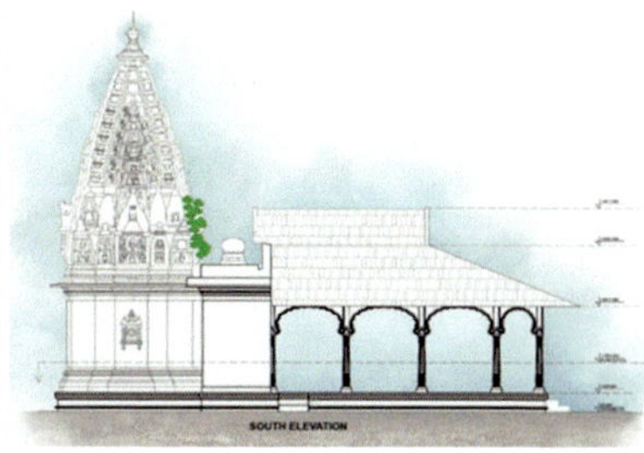 Mandapas in Nagpur have circular based two rowed columns and followed the same throughout Nagpur
 The columns in other regions have different stylistic detail like the springing line of the arch here have Kel-phul (Banana flower) as detail	 The columns in Nagpur have Peacock on the springing line of the arch and the arch depicts the tail of that peacock

Fig. 14 Timber column in Murlidhar temple. (*Source* Author)

Fig. 15 Section of Murlidhar temple. (*Source* Author)

Fig. 16 Mandapa of Rukmini temple. (*Source* Author)

eighteenth century were very unique than the other regions in Maharashtra which followed a similar style. The comparison of features showing the stylistic similarities between the temples of Bengal and Orissa and temples of Vidarbha is shown in Table 2.

4 Discussion and Conclusion

From the above analysis of temples, we understand that the temples in Nagpur constructed by Hingnikar Bhonsales in the eighteenth century were similar to the temples in other regions of Maharashtra in the nature of planning and material used in sabhamandapa. The temples in Nagpur followed the planning and construction detail of other regions in Maharashtra but the stylistic details such as carvings on the temple walls, design features on shikhara and iconography was inspired from the temple styles of Orissa and Bengal.

The war activities and rule of Hingnikar Bhonsales over the Orissa and parts of Bengal as mentioned above in the methodology suggest the prolonged stay and activities of Bhonsales in the region. This might have resulted in the influence of this region over Nagpur temple style. As mentioned above Raghuji Bhonsale was patronage of temple construction in Nagpur and many temples were constructed during his reign along with his raids on Bengal and Orissa, thus we can conclude that Nagpur temple architecture is very unique in nature as it has evolved from many influences from various regions in India.

Table 2 Comparative analysis of temples of Vidarbha, Orissa and Bengal

Temples in Bengal	Temples in Nagpur
 Terracotta temples which are unique to the Bengal have a curved sloping roof	 This curved sloping roof can be seen adopted in the Nagpur temples in the shikhara
Temples in Orissa	Temples in Nagpur
 Temples in Orissa are known for their exquisite carving on the temple walls	 We can find similar kind of carving in the temples in Nagpur
 The iconography on the walls of the temples of Orissa	 The details found in Nagpur are little bit different like the iconography on the temple walls. The sculptures are of Marathi men and women

References

Apte BK. Bhonsales of Nagpur (Chapter 7). https://gazetteers.maharashtra.gov.in/cultural.maharashtra.gov.in/english/gazetteer/History%20Part/History_III/chapter_7.pdf

Gupte KR, Oak WD (2016) Nagpurkar Bhosalyanchi Bakhar. Varada Prakashan Pvt. Ltd. ISBN-13 978-8194276753

Maharashtra State Gazetteers: Bhandara (1979) Director of government printing, stationery and publications, Maharashtra State

Human Perception Analysis of Adapting Vernacular Architectural Solutions in Urban Single Residences

Vijayalaxmi J. and Sayak Banerjee

Abstract Vernacular buildings and traditional knowledge about sustainability are getting extinct due to the global paradigm shift towards energy intensive building materials and design practices. The study focuses to identify and overcome shortcomings of traditional dwellings as it is so that it can be adopted as a sustainable housing solution for urban dwellers. Primary survey was conducted on urban residents through discussions and structured questionnaires which identified requirements from their houses in the context of modern lifestyle. This paper tries to provide solutions to overcome the limitations as perceived by the urban residents in order to increase the acceptability of vernacular considerations. The survey results indicate a lack of awareness about contemporary vernacular architecture among urban dwellers. It further explains the possibilities of adopting vernacular materials and techniques for a modern sustainable house while sufficing the requirements of urban lifestyle. Based on this study, design guidelines can be formulated for the new building constructions under various government housing schemes. The uniqueness of this study is that the bridge between existing neo-vernacular buildings and human perception has been proposed with sustainable solutions which will satisfy the client and the designer. This study will also bring an awareness among people about vernacular architecture and as they slowly start adapting the same we as a society, can regain the lost, knowledge, craftsmanship and identity of a place and head towards a sustainable society.

Keywords Adaptation · Neo-vernacular architecture · Single residence · Sustainability · Urban dwelling

---

Vijayalaxmi J.
School of Planning and Architecture, Vijayawada, India
e-mail: vijayalaxmij@spav.ac.in

S. Banerjee (✉)
Sr. Green Building Analyst Environmental Design Solutions, New Delhi, India
e-mail: sayakbolchi@gmail.com

Vijayalaxmi J. (ed.), *Conservation of Built Heritage in India*, Advances in 21st Century Human Settlements, https://doi.org/10.1007/978-981-99-8335-3_7

1 Introduction

The current era is marked by urbanization and globalization, and hence, there is a modernization of materials and construction techniques and such changes are not compatible with the surroundings. This depicts a paradigm shift in the construction industry and design practices. Extensive use of high energy materials such as concrete, glass, cement, ceramics, metal etc. combined with the increased amount of construction activity is causing an explosion in the emission levels of CO_2 gas which is eventually causing global warming (Lall 2007).

Although vernacular architecture is the reflection of the local factors of the place but it also changes as per times and lifestyle of the people to maintain the relationship between the place and the surroundings as it is termed as the identity. This identity is reduced because of the advent in globalization by using foreign material. With the changing environment, vernacular architecture has evolved to keep pace with the changing life style of people by introducing better equipment and better craftsmanship (Jagatramka et al. 2021).

Vernacular Architecture is a part of sustainability and is responsible to the local factors i.e., people and available resources. Yet it has proved to be successful in maintaining harmony with the nature. (Salman 2018). Current awareness about the environment has been an eye-opener to the people, to achieve the identity of vernacular architecture using local materials and forms, adapting vernacular principles in modern architecture in a relevant manner thereby developing the new style of architecture known as "Neo-vernacular architecture".

Contrary to the usage of industrially created materials, using traditional construction methods and materials will have various advantages (Lall 2007). Reasons for the change in vernacular architecture and indicators of sustainable architecture have been studied through case studies and comparative analysis (Jagatramka et al. 2021). Salman (2018) has discussed that the use of vernacular principles in modern construction can improve identity and sustainability of a space (Jagatramka et al. 2021). Vedamuthu et al. have discussed adaptive reuse of a vernacular house with an example of old Chettinad house. The acceptability of these buildings among urban dwellers is not discussed. The research aims to focus on the principles or elements that can be adopted from a vernacular house while constructing a new house in an urban context. It also aims to identify the aspirations of an urban dweller which need to be essentially incorporated into a vernacular house to make it acceptable for an urban lifestyle (Vedamuthu et al. 2014).

Hence this study will:

1. Help architects to propose a sustainable solution for small scale residences using local materials and techniques while achieving adequate thermal comfort in all climatic conditions.
2. Help to revive the identity of a place and also the skills of the craftsmen by the use of traditional knowledge and skill which is getting lost due to globalization.

3. People from lower economic strata follow people from higher economic strata. Therefore if there is enough awareness among the people of higher economic strata it will be easier in the paradigm shift of aspiration.

2 Study Area

Since the independence of India, the urban population growth has been critical due to various reasons which include migration of rural people to cities for better employment opportunities. As the city grows, there are a number of houses which sprawl up in the suburban area of a city, which eventually get integrated into the city, as its boundaries grow with time. These houses are built on plotted development of private developers or Government bodies like the Housing Board. Hence the study area is a suburban area with growing independent houses. As the city grows, the Government Housing bodies and private developers focus on the growth in suburban areas. These plotted developments for MIG and LIG populations are considered for study.

Secondarily three case studies of buildings have been chosen from Pathanamthitta, Kerala and Bangalore, Karnataka which has Warm, humid and temperate climate respectively. All three buildings are situated in suburban areas and fulfills the purpose of the study as they are designed and built in a neo-vernacular style of architecture.

3 Methodology

The design of the study is based on a qualitative approach where the result is arrived after comparing primary survey with secondary studies from internet and books as shown in Fig. 1. The primary study is based on structured interviews, discussions and questionnaire surveys of LIG and MIG residents of urban and suburban cities through informed consent. The participants of the study were briefed about the study before the interview and they are expected to discuss their understanding about vernacular architecture and its use in modern day buildings. Furthermore, the questionnaire seeks to understand the shortcomings of vernacular architecture as perceived by common people. Secondarily 3 buildings of neo-vernacular style from different places have been studied and inferences have been drawn to provide solutions to the issues identified through the questionnaire to generate an integrated design guideline.

3.1 Practical Implementation of the Study

This study will help common people to be aware of alternative building materials and techniques which are capable of serving contemporary needs other than the global material used in everyday construction. Vice versa it will help architects to understand

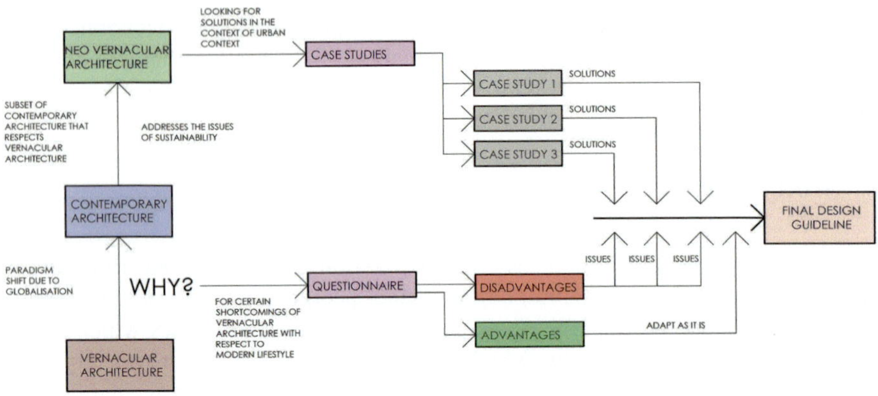

Fig. 1 Methodology of the study. (*Source* Author)

the aspiration of common people to be accommodated in the residence. Furthermore, these kinds of buildings will be more acceptable among common people if the government showcases these constructions under Pradhan Mantri Awas Yojana (PMYA-U-HFA) through platforms like ASHA–India (Affordable Sustainable Housing Accelerators) and GHTC-India (Global Housing Technology Challenge). As India's population increases more and more people are settling at the outskirts of the cities. We can regain the lost identity of a place by strategic intervention of vernacular buildings.

3.2 *Case Studies*

Three different single family residences have been chosen as shown in Table 1 which are relevant to the topic for desktop study. The following buildings have been designed and constructed keeping the sustainability aspect in mind such as traditional planning, use of local material and construction techniques, use of waste and low energy material etc. Simultaneously they also show the efforts of well-trained professionals to accommodate modern needs through conscious technicality, therefore these buildings have characteristics of neo-vernacular architecture to some extent. These structures are built in contemporary times, showing the immortality of vernacular architecture. Likewise, these buildings are located in different climatic zones exhibiting the various methods of adapting vernacular angles into a structure. Each contextual analysis may outline comparative qualities of neo-vernacular architecture in alternate ways yet they all will in general serve a single need.

Analysis of these buildings will help to establish a formwork of neo-vernacular characteristics which can be compared with the design brief formulated from the feedback of people.

Table 1 Details of buildings under case study

No	Building name	Architect	Location	Objective of study
1	Biju Mathew's residence (Debris House, Wallmakers 2018; Daniel n.d)	Ar. Vinu Daniel	Pathanamthitta, Kerala	Understanding the approach of new building construction with waste and local material
2	House of five elements (https://tropicalbuildings.org/case_studies/67; Manjunath; https://worldarchitecture.org/architecture-projects/gzce/house-of-five-elements-project)	Mansaram architects	Bangalore, Karnataka	Understanding the approach of energy neutrality by use of local material in a traditional Kannada home
3	Malhar-Resonance (http://www.manasaramarchitects.com/projects/fiveElements.php; Goodearth)	Good earth organization	Bangalore, Karnataka	Understanding the design parameters for acceptability of a neo-vernacular building by mass population when portrayed as luxury

4 Results

47 responses were collected from common people of different ages and backgrounds through structured questionnaires out of which 3 were incomplete and 1 was repetition. The following conclusions have been made based on the remaining 43 responses. This questionnaire intends to find out if common people are aware of vernacular materials and principles as a part of modern building construction, whether they are willing to accept it as an alternative to globally used materials and techniques for a sustainable built environment, if yes then what are the requirements of urban lifestyle that the neo-vernacular home has to offer.

The survey shows that 86.4% of the population wants to build their houses either within the city (43.2%) or at the outskirts of the city (43.2%) for connectivity and surrounding environment respectively (Figs. 2 and 3). Therefore this group of people represents the target population of urban dwellers.

The survey result further shows that 82.9% of the participants live in houses built with conventional materials which are globally available and energy intensive. The reason for this is the development of modern building material, availability of the same and lack of traditional knowledge and awareness among people. Although 54.8% people are aware of contemporary vernacular building only 38.1% wants to live in such a house. Reason for this kind of a distribution is identified by the structures questionnaire in the following content (Figs. 4, 5 and 6).

A lot of people actually look at vernacular architecture as an eco-friendly alternative (Fig. 7) but a group of people also look at it as a house for poor people

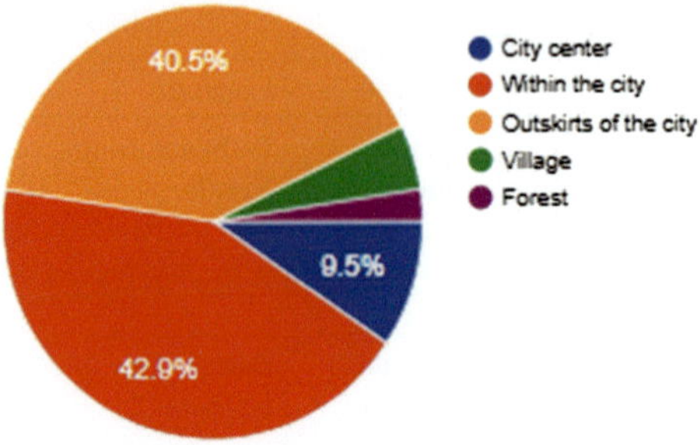

Fig. 2 Preferred location for house construction

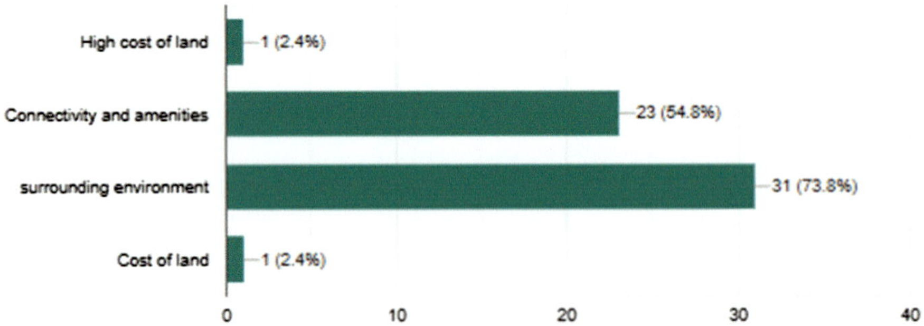

Fig. 3 Parameters for choosing location for building house. (*Source* Author)

Fig. 4 Current building typology

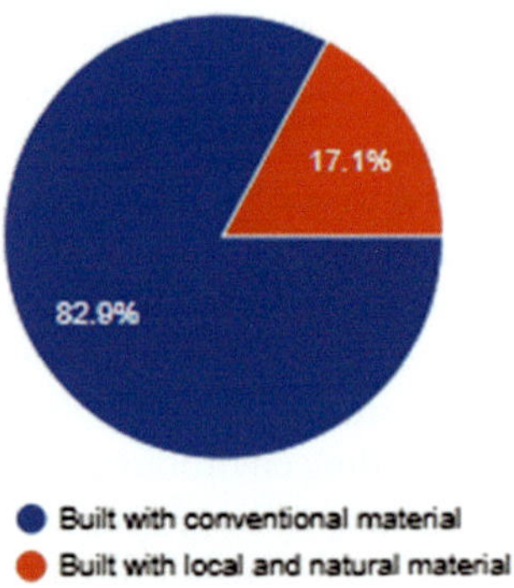

(Fig. 7) therefore when asked about the relation between vernacular buildings and economic status people tend to join dots as shown in Fig. 8. This shows the mindset of the people towards vernacular buildings which becomes a barrier when it comes to building own house in the same manner. There are other reasons which have come up for both liking and disliking vernacular buildings as shown in Figs. 9 and 10.

Through this study we can understand that majority of people are concerned about structural stability and maintenance of vernacular buildings (Fig. 11). Through some

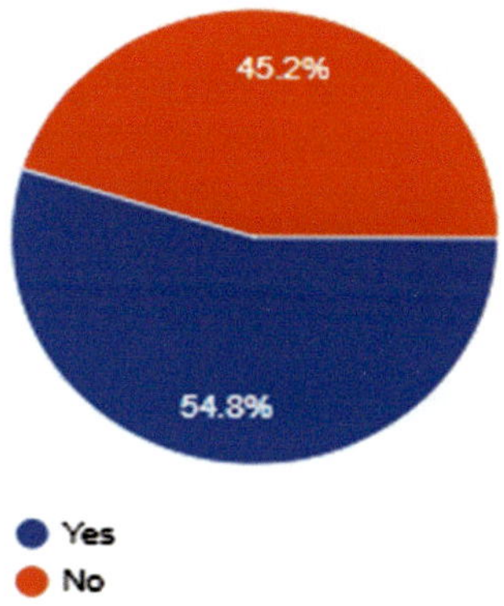

Fig. 5 People awareness about contemporary vernacular building

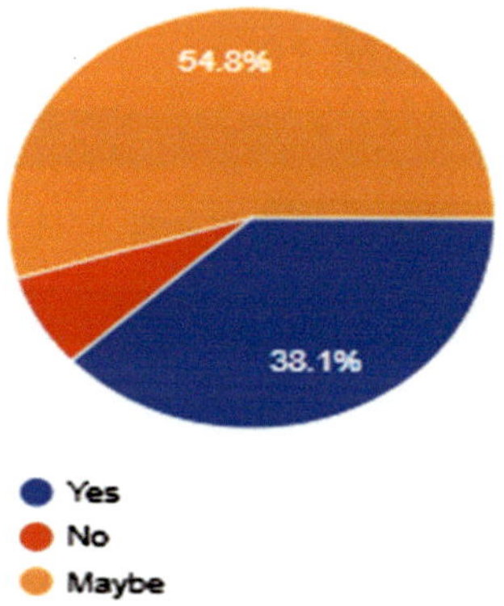

Fig. 6 People willing to live in a contemporary vernacular building. (*Source* Author)

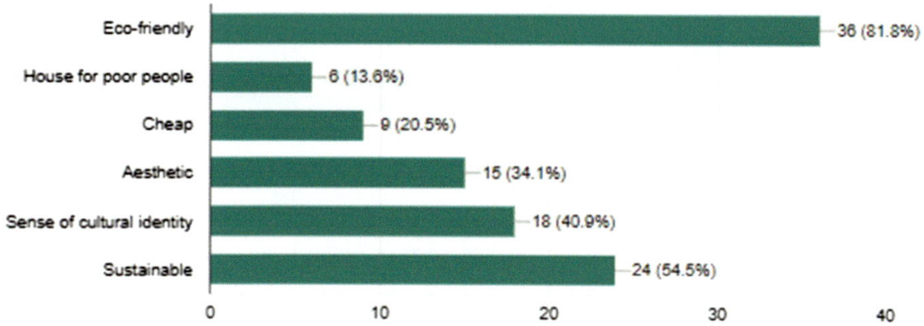

Fig. 7 Pre conceived thoughts about vernacular building

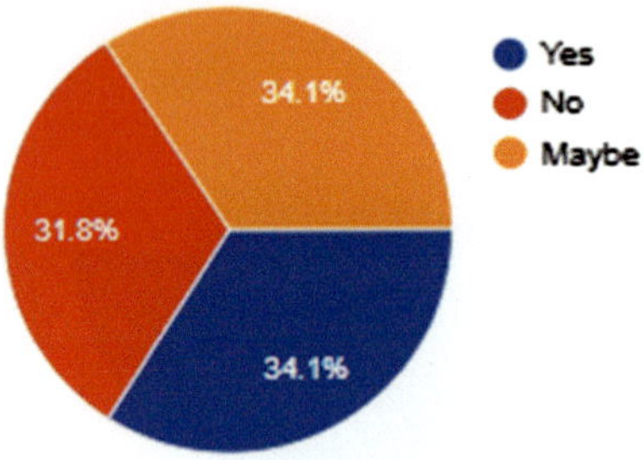

Fig. 8 Connection between economic status and buildings built with natural or waste material. (*Source* Author)

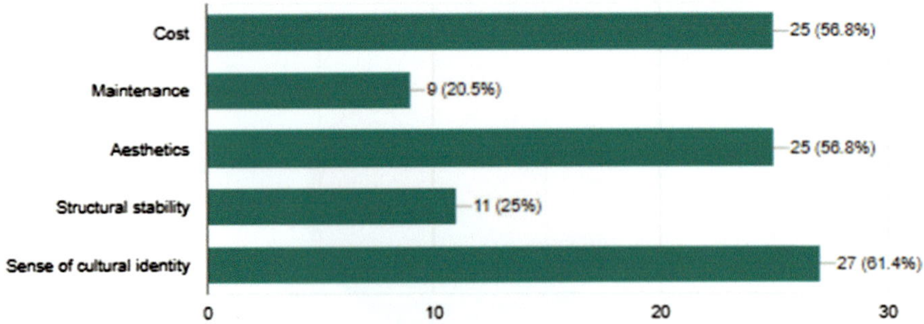

Fig. 9 Reasons to build vernacular building

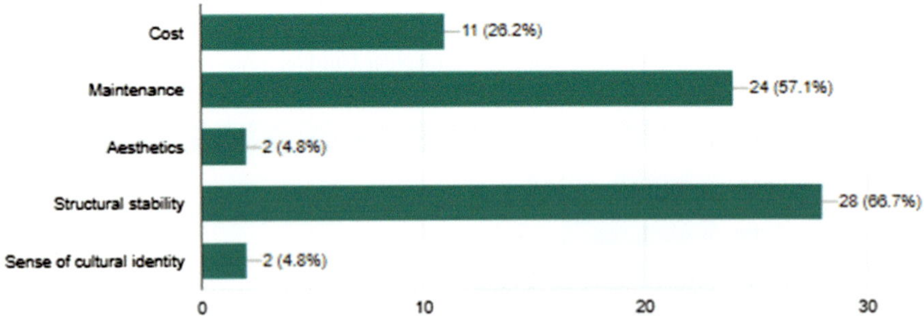

Fig. 10 Reasons not to build vernacular building. (*Source* Author)

other discussions it was evident that availability of masons with proper knowledge is another major aspect.

The survey result further focuses on the requirement of amenities for an urban lifestyle which includes some basic and some special features. We can see that majority of people opt for natural means for thermal comfort (Fig. 12) but the design

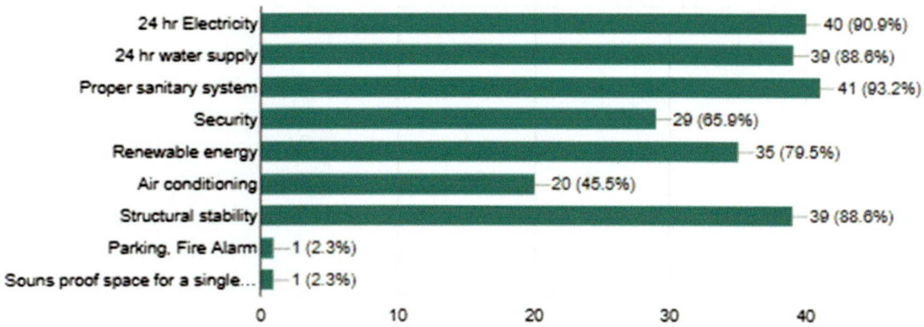

Fig. 11 Amenities required in an urban house

of the building is not efficient in most of the cases therefore the residents end up using mechanical means (Fig. 13).

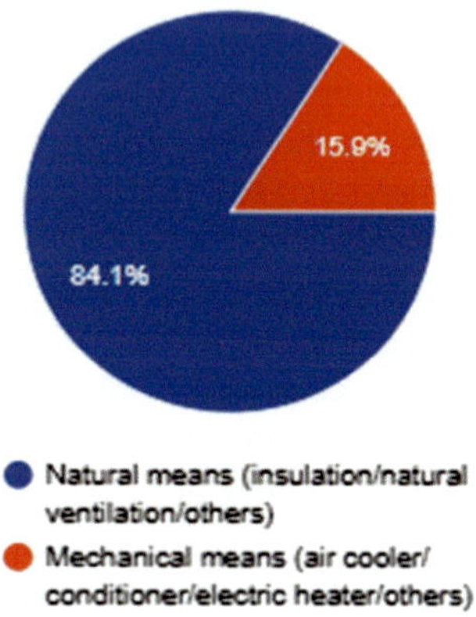

Fig. 12 Requirement of mechanical thermal comfort systems. (*Source* Author)

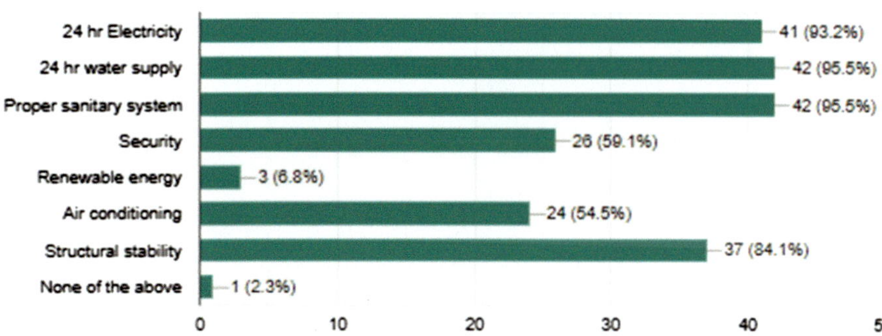

Fig. 13 Amenities presently available in houses. (*Source* Author)

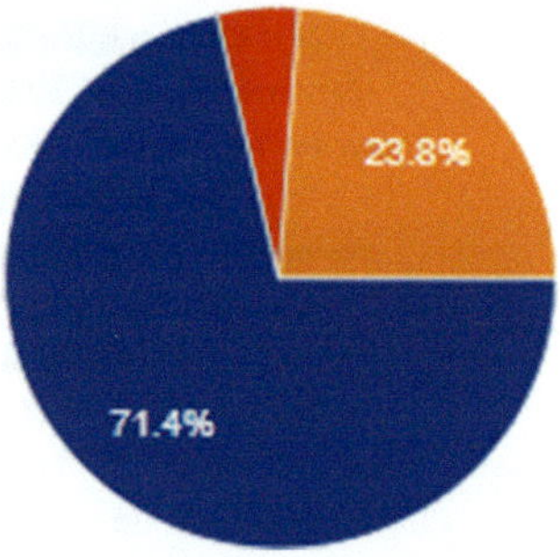

Fig. 14 Willingness to build sustainable building using vernacular principle. (*Source* Author)

When asked if they want to contribute to a healthy society by building sustainable buildings 71.4% people have answered positively (Fig. 14). This portrays the interest of common people towards sustainability but the lack of knowledge about sustainable construction. With a little guidance by experts, this interest can be utilized and channelized towards building neo-vernacular buildings with modern interventions.

In the following section, a framework is proposed (Table 2) where the issues of the vernacular house as perceived by common people have been identified. Along with that the basic requirement that the house needs to serve for an urban lifestyle has also been listed down. Further, it has been demonstrated how the buildings under study have handled those issues to provide a sustainable solution which is capable of satisfying the urban needs. The solutions shown in the table are only one way to deal with the issue. The issues can be solved for a particular issue in a particular place following the same principle of problem solving. Therefore the solutions are more of a generic solution which can change with the change of geographic location, climate, socio cultural behavior, livelihood etc.

Table 2 Framework for solving issues of vernacular architecture and urban requirement

Sl. no	Observations/ Issues/ Requirements	Solution from the case studies	Reference project	Pictorial Representation
01.	Availability of local/waste material	1. Remnant of the previous structure has been used for central curved wall as debris wall. Other walls are made with rammed earth where the soil is taken from site. Wood and other elements from junkyard has been used for interiors. Coconut shells used for filler slabs. 2. CSEB made from soil dug for water body used for walls and filler slab; Extensive use of bamboo for walling, Flooring, roofing and furniture which is locally available; finishing with natural stones, lime and certified wood. 3. Stabilized earth block, local stone, reclaimed timber, locally made terracotta tiles.	1. Biju Mathew house 2. House of five elements 3. Malhar Resonance	  
02.	Availability of knowledgeable mason	1. This building got built with unskilled labors working under proper supervision of architects.	1. House of five elements	-
03.	Structural stability	1. Debris wall compressive strength 3.5-3.78 Mpa; Rammed earth wall compressive strength 5-20 Mpa; Ferro cement vault load carrying capacity 1200 Kg/m2 2. CSEB compressive strength 4-6 Mpa; Bamboocrete used for walls and roof were tested on site for structural stability.	1. Biju Mathew house 2. House of five elements	 
04.	Maintenance	1. Floors, walls and ceiling have a rustic finish so minimum maintenance is required. 2. Proper maintenance of the bamboo and other natural materials are required.	1. Biju Mathew house 2. House of five elements	
05.	Thermal comfort measures	1. Central courtyard along with large openings allow proper ventilation in the warm humid climate; thick rammed earth walls have lower U value with higher thermal lag; coconut shells in the filler slab gives additional insulation 2. Ventilators on every window and door guarantee cross ventilation of spaces; utilization of earthenware, filler slabs, mud blocks and stone in the structure keeps the insides cool; the presence of an internal yard with a water body keeps up humidity and improves the thermal comfort levels.	1. Biju Mathew house 2. House of five elements	 
06.	Building cost	1. Building cost is comparatively less : 8827 rs/m2 2. Price of the apartment : 1.75-2.45 cr (including land cost)	1. House of five elements 2. Malhar Resonance	-
07.	Conflict with socio economic status	1. The house is situated in an urban community in a city like Bangalore where people of higher economic strata resides. 2. The housing community involves a varied blend of occupants – from the craftsman, software engineers to the architects.	1. House of five elements 2. Malhar Resonance	-
08.	Security	1. Grills made of scrap meter boxes. 2. Layout is secured with a solid 6 feet compound wall.	1. Biju Mathew house 2. Malhar Resonance	 
09.	Proper sanitary system	1. Proper plumbing system for water supply and waste with PVC pipes; all modern facilities available in the washrooms.	1. Malhar Resonance	-

Sl no	Observations/ Issues/ Requirements	Solution from the case studies	Reference project	Pictorial Representation
10.	24 hr electricity	1. All light fixtures are either CFL or LED and the electricity is generated through roof top solar panels with a capacity of 1KW/day. 2. Minimum lighting for streets and parks and CFL and LED for houses; Low wattage equipment for water heating.	1. House of five elements 2. Malhar Resonance	
11.	24 hr water supply	1. Rain water harvesting in sump of capacity of 75000 ltrs and DEWAT system for waste water treatment which purifies 500 ltrs /day 2. Underground rainwater harvesting tank and recharge pit for excess water; sewage water treatment plant with bio filters for entire community.	1. House of five elements 2. Malhar Resonance	
12.	Renewable energy	1. Solar panels with a capacity of 1KW/day.	1. House of five elements	-
13.	Parking	1. Car porch has been provided. 2. 2 nos of open car parks have been provided in each villa.	1. Biju Mathew house 2. Malhar Resonance	-
14.	Fire safety	1. No proper fire safety measures have been taken, but in general the materials used for walls like rammed earth and CSEB walls are fire resistant.	1. All projects	
15.	Inspiration from traditional houses	1. Central courtyard as an traditional Kerala house bringing the sense of identity 2. Central courtyard with water body as an traditional "Thotti mane" style house 3. Open to sky garden, suggestive of the "Aangans" of traditional homes.	1. Biju Mathew house 2. House of five elements 3. Malhar Resonance	 
16.	Awareness about sustainability	1. The project involves local people and younger students of the encompassing areas.	1. House of five elements	-

5 Discussion

When we shifted to construction with global materials as a result of industrial revolution, the building designs also started becoming more and more constrained and modular and at the same time lost its uniqueness or identity. Brick became the common construction material instead of mud, stone or bamboo. As the cities started growing with these energy intensive materials slowly the question of sustainability arose. Many of the previous researchers have tried to look for sustainability solutions in traditional architecture. These solutions have been given based on case studies, surveys, simulation and mathematical calculations of the traditional dwellings. In many of the cases, the solutions proposed are as simple as orientation, form or building material. Unfortunately, very few of them have assessed the acceptability of those solutions by the mass population. Standing in an era of modernization house has become a status symbol for middle and lower economic strata of society. Most of the people deny to avail those solutions due to various reasons including scarcity of land, non-availability of material and skilled labor or even the aesthetic and maintenance point of view. In this situation, this paper intends to bridge the gap by bringing out the requirements of people from their housing and discussing how the solutions can be made more adaptable.

It is evident that neo-vernacular houses as shown in the case studies have been able to solve many of the issues related to vernacular buildings to improve the acceptability. Now to increase its spread both central government and state government will be useful. There are numerous housing projects promoted by various state and central governments which include schemes for HIG, MIG, LIG and EWS. There are options for individual villas with plots in the suburban and peri-urban areas and apartments as well in urban areas. As this paper mainly focuses on solutions to single family residences so, for example, there is area development scheme for 520 (HIG-85, MIG-151, LIG-40 & EWS-244) plots at Villankuruchi village, Coimbatore north taluk, Coimbatore district under Tamil Nadu housing board or Shilpakanan phase I, phase II and phase III in Durgapur under West Bengal housing board. Vernacular architecture of those specific locations can be applied in these projects along with the solutions proposed then it will have better reach to people and also be a more sustainable solution to the housing requirement of the country.

As a suggestion for future research more number of neo-vernacular houses can be studied to suggest acceptable solutions across different climatic zones of India. Further studies can be carried out to look for solutions in mass housing to expand the acceptability of the solutions given to common people.

6 Conclusion

Vernacular architecture has grown up at a particular place to cater to their specific needs of dwelling with the help of limited resources available locally. Due to globalization we are about to lose our indigenous knowledge about sustainable construction and living. But it is still not too late as vernacular architecture is still practiced in many of the country sides thus a lot of people associate vernacular architecture with a particular section of society. This type of architecture resonates less with people looking for design solutions to meet their urban requirements. The aspiration of the lower socio economical group plays a big role here. Although few architects are trying to blend contemporary architecture with vernacular architecture to give birth to a new style of architecture which at the same time serves the aspiration of the urban dwellers while respecting vernacular principles. If the city dwellers start being vocal about local materials and techniques that can set new trends for the rural population for whom the aspirational quotient comes from urban population. With the recent outcry about sustainability, it is the best time to get back to our time tested vernacular architecture for the urban population. Same can be achieved by promoting sustainability at a smaller scale and at an affordable cost for all categories of people.

References

Daniel V (n.d) Biju Mathew house, Pathanamthitta wallmakers. www.wallmakers.org. Accessed 17 June 2021

Debris House/Wallmakers (2018) ArchDaily. ISSN 0719–8884. https://www.archdaily.com/903691/debris-house-wallmakers. Accessed 17 Jun 2021

Goodearth, Malhar Resonance Brochure

http://www.manasaramarchitects.com/projects/fiveElements.php

https://tropicalbuildings.org/case_studies/67

https://worldarchitecture.org/architecture-projects/gzce/house-of-five-elements-project

Jagatramka R, Kumar A, Pipralia S (2020) Sustainability indicators for vernacular architecture in India, ISVS E-J 7(4)

Jagatramka R, Kumar A, Pipralia S (2021) Transformations of vernacular architecture of India: problems and prospects. ISVS E-J 8(1)

Lall AB (2007) Evolving traditional practices for sustainable construction in the present. In: INTBAU International Conference, 2007. https://www.scribd.com/document/468988707/Evolving-traditional-practices-for-sustainable-construction

Manjunath N Symphony of the Bamboos. In: XI World Bamboo Congress, Xalapa, Mexico

Salman M (2018) Sustainability and vernacular architecture: rethinking what identity is. In: Hmood K (ed) Urban and architectural heritage conservation within sustainability. https://doi.org/10.5772/intechopen.82025

Vedamuthu R, Dhakshinamoorthi J, Sasidhar K (2014) Adaptive reuse and restoration of a Chettinadu Mansion, Pudukkottai, Tamilnadu, India. In: SAHC2014—9th International conference on structural analysis of historical constructions

Public Perception of the Significance of Cultural Heritage Conservation for Promoting Community Wellbeing

Manali Basu

Abstract Target 11.4 of the Sustainable Development Goals (SDGs) addresses the fact that efforts should be made to protect and safeguard the cultural and natural heritage of the world. Cultural heritage needs to be protected and conserved as it helps in preserving the unique identity and character of a place. In order to achieve this, assessing the perceptions of the local community about the conservation of built heritage becomes an important aspect, as a positive perception means that the conservation efforts will be a success. The research is conducted to assess the public perception of conserving the cultural heritage of a place and its role in promoting community wellbeing. Purposive sampling method through informed interviews is used on the participants. Also, a questionnaire prepared by using the Semantic Differential Scale (SDS) and Likert Scale was used to conduct the survey. The data is collected with the help of a questionnaire survey which is carried out by taking a sample consisting of 106 teachers from the schools and colleges in Amaravati and Vijayawada, Andhra Pradesh, India. The results are analysed using MS Excel software. The results show a positive perception and awareness among both the male and the female teachers regarding their local cultural heritage and they want to take part in the conservation process by promoting their cultural heritage throughout the world. Conserving the heritage by balancing it with other sustainability goals in community planning can help in developing cities that are sustainable with respect to the three pillars of sustainability.

Keywords Community wellbeing · Cultural heritage · Heritage conservation · Public perceptions · Semantic differential scale

M. Basu (✉)
School of Planning and Architecture, Vijayawada, India
e-mail: basumanali94@gmail.com

Vijayalaxmi J. (ed.), *Conservation of Built Heritage in India*, Advances in 21st Century Human Settlements, https://doi.org/10.1007/978-981-99-8335-3_8

1 Introduction

On a broader scale, heritage can be defined as almost everything that humans have experienced or created since the time they have gained some consciousness. According to the 2009 UNESCO Framework for Cultural Statistics, cultural heritage consists of artefacts, monuments, a group of buildings and sites and museums that have a diversity of values including symbolic, historic, artistic, aesthetic, ethnological or anthropological, scientific and social significance. It also includes the intangible cultural heritage (ICH) or living heritage which comprises of living expressions that have been inherited from our ancestors, such as oral traditions, performing arts, social manners, rituals, festive events, knowledge and practices related to nature and the universe, and knowledge and techniques linked to traditional crafts, that act as an important factor in maintaining cultural diversity. Built heritage forms an integral part of this cultural heritage. It consists of the various historical structures that have been created throughout the world since the very beginning of the human civilization that act as significant locations of human experience even in the present day (Taçon and Baker 2019).

On the other hand, wellbeing may be defined in many ways but the most appropriate definition would be a feeling of content, comfort and happiness which is generally associated with the physical, mental and emotional health. From the empirical researches conducted in the historic cities by the researchers with different disciplinary backgrounds, it was found that meaningful encounters of people with various types of heritage can have a positive impact on the physical and mental peace of the perceivers which leads to an enhancement of individual and community wellbeing. Whereas, damages to the heritage can lead to a demoralizing effect that will lower the wellbeing (Taçon and Baker 2019). The concept of community wellbeing falls in between the individual wellbeing and the wellbeing of the nation which potentially addresses the concept of 'being well together' which is generally neglected. Community wellbeing can be defined as 'the combination of social, economic, environmental, cultural, and political conditions identified by individuals and their communities as essential for them to flourish and fulfil their potential' (Wiseman and Brashar 2008).

Cultural heritage needs to be conserved as it is an asset that protects the past for the present and the future generations and can contribute to the three pillars of sustainability by creating an awareness for limiting the use of resources, by increasing employment opportunities, promoting community as well as individual wellbeing and a sense of social inclusion. Further, it helps in preserving the local identity that gives a sense of belonging and pride to the local people (Lerario 2022) which is gradually getting lost due to globalisation, increased urbanisation and over-tourism.

1.1 Review of Literature

When it comes to cultural heritage conservation, assessing the awareness and perceptions of the local community becomes an important aspect, as it has a direct influence on the heritage conservation and interventions (Shankar and Swamy 2013; Srivastava 2015; Sektani et al. 2021). When considering the contribution of the cultural heritage to the wellbeing specifically, it has to be understood that it is a multifaceted social construct that has the potential of delivering benefits to the people and the entire society which is commonly known as the instrumental values (Ateca-Amestoy et al. 2021).

Regarding awareness about the cultural heritage among the local public, Srivastava (2015) in her study with the teachers of Dayalbagh Educational Institute (Deemed University) in Dayalbagh, Agra, UP, found that though both the male and the female teachers were aware of the cultural heritage but the overall score of the female teachers were better than the males in almost all the cultural heritage dimensions showing that they are more aware. Shimray and Ramaiah (2019) found that the male and the female students of the Pondicherry University do not differ in their awareness regarding the cultural heritage, acquiring cultural knowledge and promotion of cultural heritage.

When it comes to wellbeing, Gallou (2022) aimed to assist the researchers in understanding the ways in which heritage can stimulate the outcomes of mental or physical health in a direct or indirect way and identify how these benefits take place at the individual as well as community level to a greater extent. Gallou et al. (2022) found that visits to heritage sites have potential therapeutic benefits for visits that are self-directed and therefore have potential means of providing support sustainably for the public wellbeing. Sektani et al. (2021) in their study, found that the results related to wellbeing of the residents in the Erbil City showed significant differences between the perceptions of the architects and the non-architects. It was found that the heritage buildings did not contribute to the non-architects' wellbeing whereas it contributed to the wellbeing of the architects. Further, it was observed that the educational background plays a major role in the perception of the built heritage. Innes et al. (2021) conducted a study in the UK with 48 people to find out the contribution of the heritage sites for promoting the wellbeing of the people suffering from dementia by taking their interviews post-session and through pre-and post-mood questionnaires. The results showed that the heritage sessions were highly valued by the participants and had positive impacts on their individual wellbeing as well as their relationship with each other.

Coming to community wellbeing, Kee and Chau (2019) investigated the economic impact related to architectural heritage in Hong Kong and found that conservation of heritage can successfully promote sustainability related to the economy apart from cultural and social sustainability. Stolare et al. (2021) in their study found that the Swedish primary school teachers supported the concept of visiting the heritage sites as this can contribute to a better historical learning process than simply teaching in the classroom.

From the review of literature, it can be observed that several studies have been carried out to find out the awareness of the people regarding the cultural heritage and on the impact of cultural or natural and/or built heritage on the individual as well as community wellbeing in different parts of the world. But in case of the India, limited studies have been found that try to find out the relationship between the cultural heritage and community wellbeing by means of understanding the public perceptions. Also, the awareness studies conducted in India involved teachers and students from single universities only. Hence, the aim of this study is to bridge this gap that has been found through the review of the literature. The present study focuses on the awareness and the perceptions of the local teachers both at the school and university levels in Amaravati and Vijayawada, as considering a diverse population of teachers is more rational (Srivastava 2015). Since, it becomes important to understand the awareness and perceptions of the local people, as a positive awareness and perception will lead to a successful restoration process, therefore, the research will aim to answer the following questions:

1. How aware the teachers are about their local cultural heritage?
2. How do they perceive the local cultural heritage in terms of visual perception?
3. How does the local cultural heritage influence their individual wellbeing?
4. How important do they feel it is to conserve the local cultural heritage for promoting community wellbeing?
5. What is their opinion about taking part in the process of local heritage conservation?

Based on the findings in the review of literature, the following hypotheses shown in Table 1 were developed:

In a country like India which is known for its rich cultural diversity and architectural heritage, it becomes very important to conserve the heritage which narrates, and will be narrating the stories of the glorious past to the present and the future generations. Further, this study will be useful to the built environment, as previous studies have mentioned that the retention of the heritage by combining it with other sustainability goals in community planning can lead to cities that will be sustainable with respect to the three pillars of sustainability.

2 Study Area

The study was conducted in the schools and colleges of Amaravati and Vijayawada in Andhra Pradesh, India. Amaravati is located along the banks of the Krishna River which is known for its rich Buddhist cultural heritage. It contains the remains of the ancient Buddhist Stupas and Mahachaitya. It is a pilgrimage site for both the Buddhists and the Hindus. It contains several temples dedicated to Lord Shiva, among which the Lord Amaralingeswara or the Amareswara Swamy Temple is the most famous one.

Table 1 Hypotheses developed for the study

Hypotheses	Statement	Parameters
H1	There is a significant difference in the awareness regarding the local cultural heritage among the male and the female teachers in terms of:	a. Historical structures
		b. Traditional handicrafts
		c. Oral traditions
		d. Performing arts
H2	There is a significant difference in the individual wellbeing of the male and female teachers regarding the local cultural heritage in terms of:	a. Happiness
		b. Satisfaction
		c. Pride
		d. Comfort
		e. Attachment
H3	There is a significant difference in the perception of the male and female teachers regarding the local cultural heritage conservation for promoting community wellbeing in terms of:	a. Maintaining the cultural context and identity of a place
		b. Process of historical learning and development of knowledge
		c. Promoting local business and job opportunities
		d. Generating local economy by drawing tourists
		e. Reducing the materials and resources consumption
		f. Reducing CO_2 emissions
		g. Promoting environmental and ecological awareness
H4	There is a significant difference in the opinion of the male and female teachers about taking part in the process of local cultural heritage conservation in terms of:	a. Willingness about participation
		b. Reason for participation
		c. Way of participation
		d. Compensation expected for participation

The city of Vijayawada lies in a level plain that is punctuated by the hills on the river Krishna and lies about 130 km southwest of Rajahmundry. It is known for its Hindu and Buddhist pilgrimage sites. Some of the historical structures in Vijayawada include the Kanaka Durga Temple, the Undavalli and the Mogalarajapuram caves and the Kondapalli fort. There is a blend of rural and urban culture in Vijayawada. The head of the village in the olden days used to organise 'Burra Kathas' which is an oral tradition of this place. Apart from that, it is also known for its performing arts which is the classical dance form of 'Kuchipudi' and the wooden toys known as the 'Kondapalli dolls' which are locally known as the 'Kondapalli bommalu.' However, it is now being considered as a 'dying craft' mainly because of the decreasing number of artisans (Figs. 1, 2 and 3).

3 Materials and Methods

The study area was chosen as Amaravati and Vijayawada in Andhra Pradesh, India. The data is collected by means of a questionnaire survey which was prepared by using the Semantic Differential Scale (SDS) and the Likert Scale.

3.1 Sample

The sample for the study consists of teachers at the school and college levels who are local to the area of Amaravati and Vijayawada. Previous studies considered samples consisting of teachers at either university level (Srivastava 2015) or school level (Stolare et al. 2021). Teachers were selected from only a single university or a single school. In the present study, teachers from three schools and two colleges have been considered, as taking a diverse population of teachers as the sample is more rational (Srivastava 2015).

3.2 Data Collection Method

For conducting the survey, a questionnaire was prepared by using the Semantic Differential Scale (SDS) and the Likert Scale. It consists of six sections. The first section deals with the 'basic information' of the respondents, the second section deals with the 'awareness regarding the local cultural heritage', the third section deals with the 'visual perception' of the respondents, the fourth section deals with the 'individual wellbeing' of the respondents, the fifth section deals with the 'importance of local heritage conservation for promoting community wellbeing' and the sixth section deals with the 'participation in the local heritage conservation'. Out of these six sections, sections three to five are measured by using the Semantic Differential

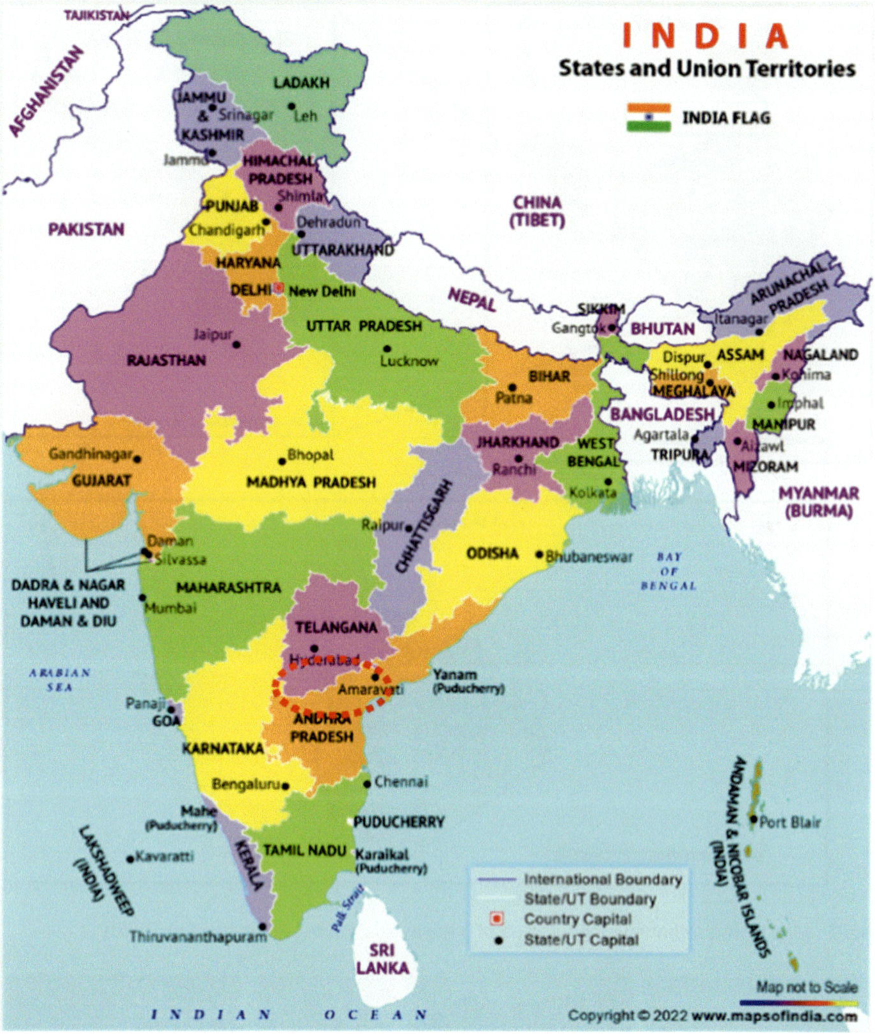

Fig. 1 Map of India showing the location of Amaravati. (*Source* Google)

Scale (SDS) and the Likert Scale. The questionnaire consists of mainly closed-ended questions but each section had an open-ended question at the end where the respondents can address any other important point(s) related to that particular section if that was not present in the closed-ended questions.

Before collecting the data, a pilot study was conducted with 50 questionnaires to test the internal consistency and reliability of the questionnaire. Cronbach's alpha was calculated by using Formula 1 for each of the three constructs shown in Table 2 and was found to be in the acceptable range that is greater than 0.70.

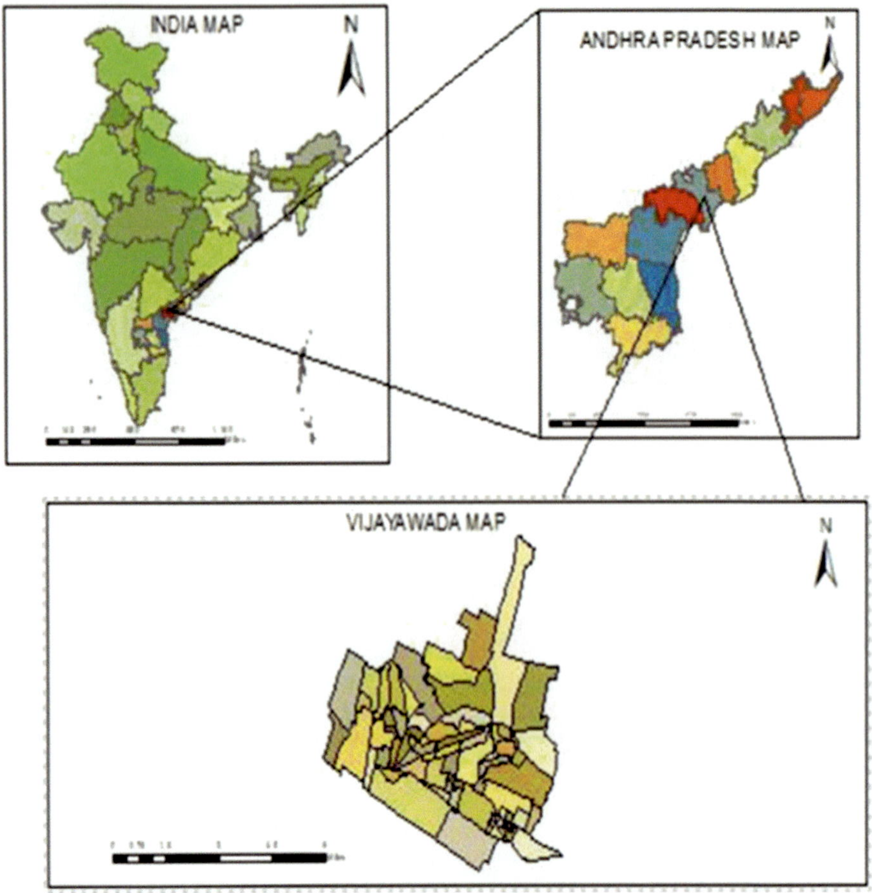

Fig. 2 Map of India showing the location of Vijayawada. (*Source* Harsha et al. 2020)

Fig. 3 Local cultural heritage of Amaravati and Vijayawada, Andhra Pradesh, India. (*Source* Google)

Table 2 Cronbach's alpha (reliability) of each of the three constructs

Section 3: Visual perception of the respondents			
Sl. no	Parameters	N	Cronbach's alpha (α)
3.1	Impression	50	0.748
3.2	Interesting	50	
3.3	Peaceful	50	
3.4	Pleasant	50	
3.5	Complexity	50	
3.6	Positivity	50	
3.7	Beauty	50	
3.8	Cleanliness	50	
Section 4: Individual wellbeing of the respondents			
Sl. no	Parameters	N	Cronbach's alpha (α)
4.1	Happiness	50	0.823
4.2	Satisfaction	50	
4.3	Pride	50	
4.4	Comfort	50	
4.5	Attachment	50	
Section 5: Importance of local cultural heritage conservation for promoting community wellbeing			
Sl. no	Parameters	N	Cronbach's alpha (α)
5.1	Maintaining the cultural context and identity of a place	50	0.795
5.2	Process of historical learning and development of knowledge	50	
5.3	Promoting local business and job opportunities	50	
5.4	Generating local economy by drawing tourists	50	
5.5	Reducing the materials and resources consumption	50	
5.6	Reducing CO2 emissions	50	
5.7	Promoting environmental and ecological awareness	50	

$$\text{Cronbach's Alpha } = \alpha = \mathrm{K/K} - 1\left[1 - \sum \mathrm{S}_{\mathrm{y}}^{2}/\mathrm{S}_{\mathrm{x}}^{2}\right] \tag{1}$$

where, K = the number of test item, $\sum S^2{}_y$ = the sum of the item variance and $S^2{}_x$ = the variance of the total score.

After testing the Cronbach's alpha for each of the constructs, the questionnaires were distributed among the teachers of the schools and colleges in Amaravati and

Fig. 4 Map showing the location of the schools in Amaravati that took part in the survey. (*Source* Google Earth Pro)

Vijayawada. Before distributing the questionnaires, permission was taken from the Principals of the respective schools and colleges. Out of the five schools shortlisted in Amaravati, Principals of three schools namely Sree Rama Krishna Hindu High (SRKHH) School, Santhinikethan School and Sri Chaitanya School agreed to take part in the survey. Out of the four colleges shortlisted in Vijayawada, the Principals of two colleges namely Nalanda Degree College and Parvathaneni Brahmayya Siddhartha College agreed to participate in the survey. Before distributing the questionnaires to the teachers, they were briefed about the intention of the survey. Data was collected for a period of 15 days from March 3, 2023, to March 17, 2023. A total of 106 respondents participated in the survey (Figs. 4 and 5).

3.3 Method of Analysis

The collected data is analysed using MS Excel. The percentages of the responses for the different parameters are calculated. The Mean and the Standard Deviation (SD) values are also calculated. Spearman's Rank Correlation is performed to find out the correlation between the qualitative aspects and the age and gender of the respondents.

Fig. 5 Map showing the location of the colleges in Vijayawada that took part in the survey. (*Source* Google Earth Pro)

Further, to test the hypotheses, Mann Whitney U Test is performed in MS Excel as the data is not normally distributed. This is a non-parametric test which serves as an alternative to the independent sample t-test and is used when the assumptions of the t-test are not fulfilled.

4 Results and Discussion

4.1 Results

After analysing the data, the following results are obtained:

4.1.1 Results Obtained by Analysing the Basic Information of the Respondents

A total of 106 respondents participated in the survey. The results show that out of the total, 44 (42%) are male teachers and 62 (58%) are female teachers. Most of the

Table 3 Results obtained by analysing the basic information of the respondents

Measure	Items	Frequency	Percentage (%)
Gender	Male	44	42
	Female	62	58
Age group (in years)	18–24	8	8
	25–34	35	33
	35–44	39	37
	45–54	17	16
	55 and above	7	7
Place	Amaravati	39	37
	Vijayawada	67	63
Number of years of stay in Amaravati/Vijayawada	18–24	41	39
	25–34	35	33
	35–44	22	21
	45- 54	6	6
	55 and above	2	2

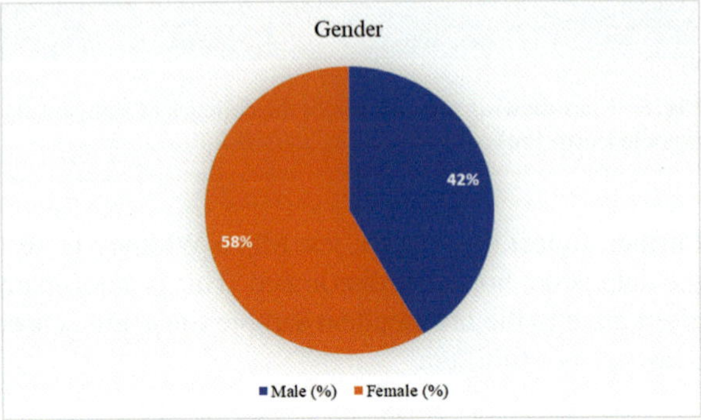

Fig. 6 Percentage of male and female

teachers (37%) belong to the age group of 35–44 years. 37% of the teachers belong to Amaravati whereas 63% belong to Vijayawada. Most of the teachers (39%) have lived in Amaravati / Vijayawada for about 18–24 years. The detailed results of the analysis have been presented in Table 3. Also, Figs. 6, 7, 8 and 9 represent the results of the analysis for basic information of the respondents.

4.1.2 Awareness Regarding the Local Cultural Heritage

Questions related to awareness about the local cultural heritage were asked to the respondents to find out how aware they are about the same. From the results shown in Table 4, it can be observed that most of the respondents are aware about their local

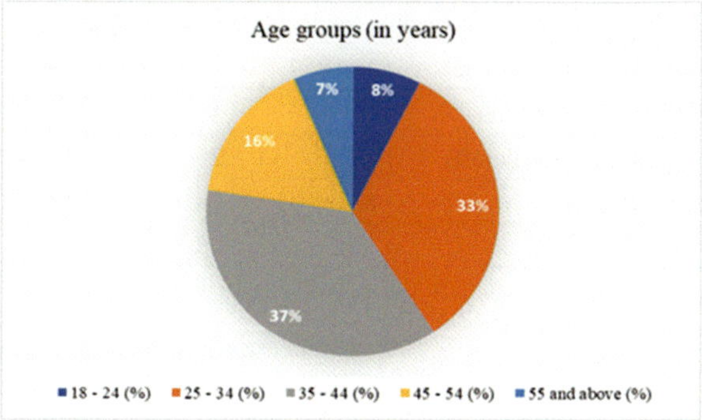

Fig. 7 Percentage of the age groups (in years)

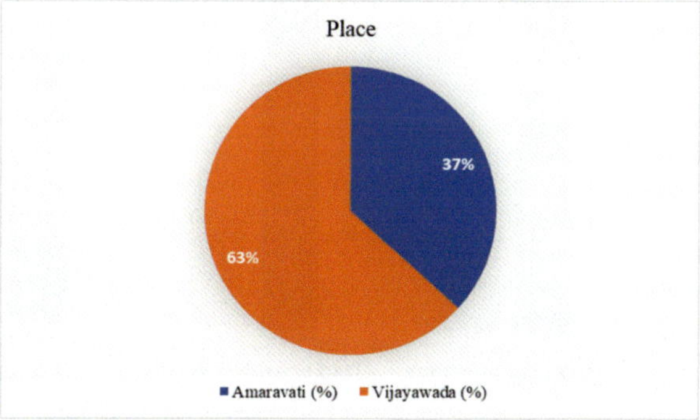

Fig. 8 Percentage of the places

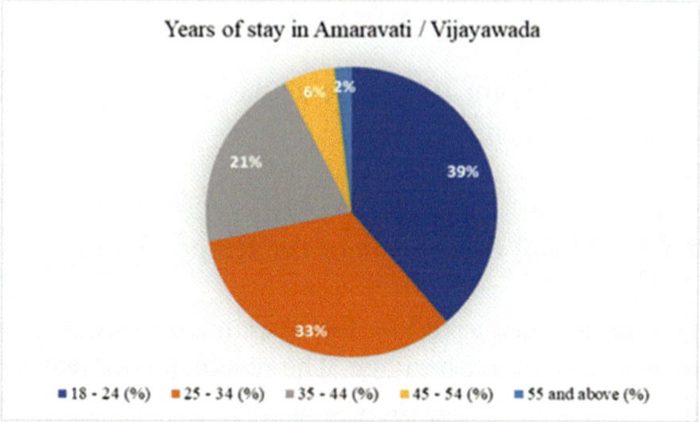

Fig. 9 Percentage of years of stay in Amaravati/ Vijayawada

cultural heritage. Most of them (82%) are aware about the 'historical structures.' The mean value is also the highest in case of the 'historical structures', supporting the fact that they are most aware about the same. The lowest awareness (62%) is about the 'oral traditions' and the mean value in this case is also the lowest (1.6226). Therefore, though most of them are aware, still some efforts should be made to strengthen the awareness so that a greater number of people become aware of their rich cultural heritage. Efforts should be made to make the people aware of their 'oral

Table 4 Awareness regarding the local cultural heritage

Local cultural heritage	Yes (%)	No (%)	Mean	Standard deviation (SD)
Historical structures	82	18	1.8208	0.3854
Traditional handicrafts	75	25	1.7547	0.4323
Oral traditions	62	38	1.6226	0.4870
Performing arts	70	30	1.6981	0.4613

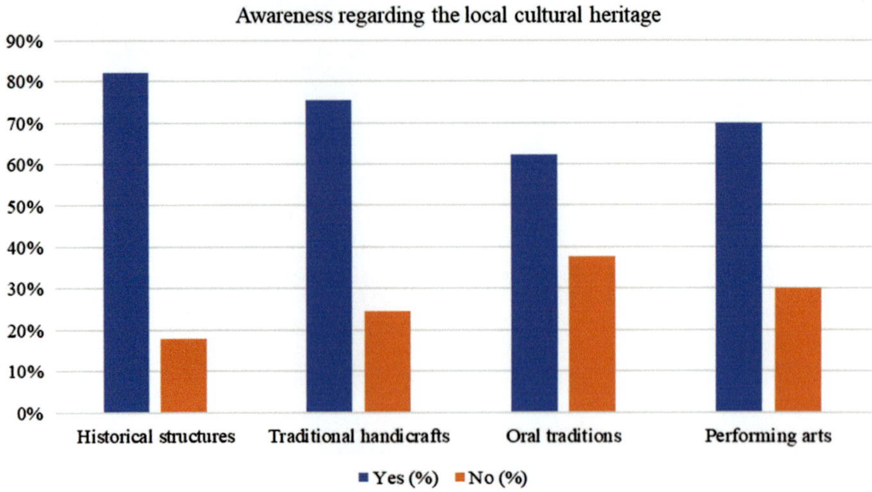

Fig. 10 Percentage of awareness among the respondents regarding the local cultural heritage

traditions.' Figure 10 represents the results of the analysis for the awareness levels of the respondents.

4.1.3 Visual Perception of the Respondents

Questions were asked to the respondents about their visual perception regarding their local cultural heritage. The results mentioned in Table 5, show that 50% of the respondents feel that their local cultural heritage is 'impressive,' 42% find it to be 'interesting,' 49% find it to be 'peaceful', 50% find it to be 'pleasant', 38% find it 'neither too simple nor too complex' that is they feel neutral about the complexity, 58% feel positive about it, 55% feel that it is 'beautiful' and 48% feel it is 'clean.' Also, it is found that out of all these eight parameters for visual perception, 'peacefulness' has the highest mean score of 4.1698 whereas, 'complexity' has the lowest mean score of 3.4057. That is most of the respondents find their local cultural heritage as 'peaceful.' Figures 11, 12, 13, 14, 15, 16, 17 and 18 represent the results of the analysis for visual perception of the respondents.

Table 5 Visual perception of the respondents

Parameter	Measure	Percentage (%)	Mean	Standard deviation (SD)
Impression	Too unimpressive	6	3.9623	0.9945
	Unimpressive	1		
	Neutral	14		
	Impressive	50		
	Very impressive	29		
Interesting	Too boring	0	4.1509	0.8257
	Boring	4		
	Neutral	16		
	Interesting	42		
	Very interesting	39		
Peacefulness	Too disruptive	1	4.1698	0.7865
	Disruptive	2		
	Neutral	12		
	Peaceful	49		
	Very peaceful	36		
Pleasantness	Too unpleasant	1	4.0566	0.8145
	Unpleasant	3		
	Neutral	16		
	Pleasant	50		
	Very pleasant	30		
Complexity	Too simple	1	3.4057	1.0026
	Simple	18		
	Neutral	38		
	Complex	26		
	Very complex	17		
Positivity	Too negative	2	3.7925	0.8014
	Negative	4		
	Neutral	22		
	Positive	58		
	Very positive	14		
Beauty	Too ugly	3	4.1226	0.8361
	Ugly	1		
	Neutral	9		
	Beautiful	55		
	Very beautiful	32		

(continued)

Table 5 (continued)

Parameter	Measure	Percentage (%)	Mean	Standard deviation (SD)
Cleanliness	Too dirty	1	4.0566	0.7908
	Dirty	1		
	Neutral	20		
	Clean	48		
	Very clean	30		

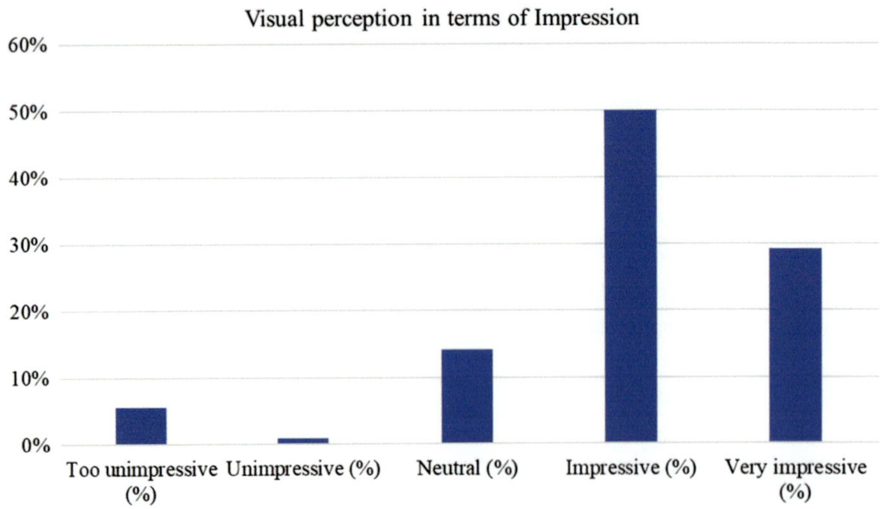

Fig. 11 Visual perception in terms of impression

4.1.4 Individual Wellbeing of the Respondents

Questions were asked to the respondents to find out the effect of the local cultural heritage on their individual wellbeing. The results in Table 6 show that 52% of the respondents feel 'happy' about their local cultural heritage, 61% are 'satisfied,' 57% feel 'proud' about it, 62% feel 'comfortable' about it, 49% of them are 'attached' while 30% are 'very attached'' with their local cultural heritage. Out of the five parameters for individual wellbeing, it is found that 'happiness' has the highest mean score of 4.1226 whereas, 'satisfaction' has the lowest mean score of 3.9623. This means that most of the respondents feel 'happy' about their local cultural heritage. Figures 19, 20, 21, 22 and 23 represent the results of the analysis for the individual wellbeing of the respondents.

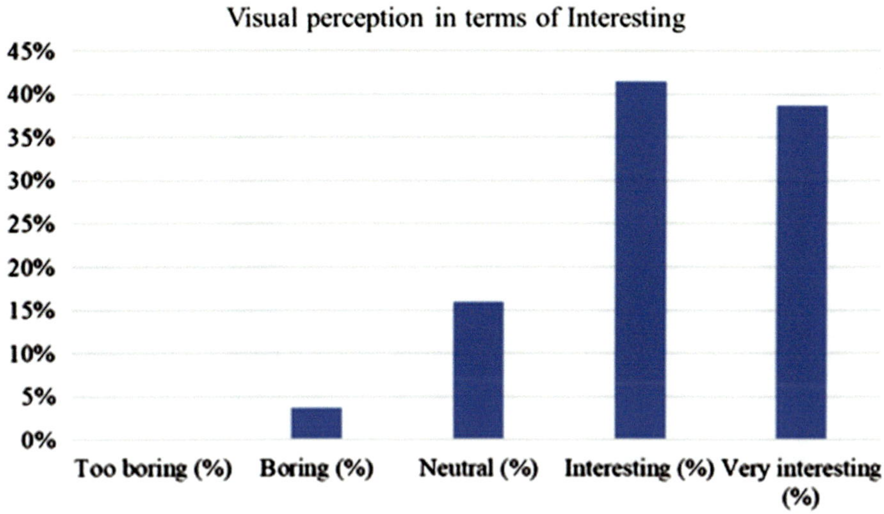

Fig. 12 Visual perception in terms of interesting

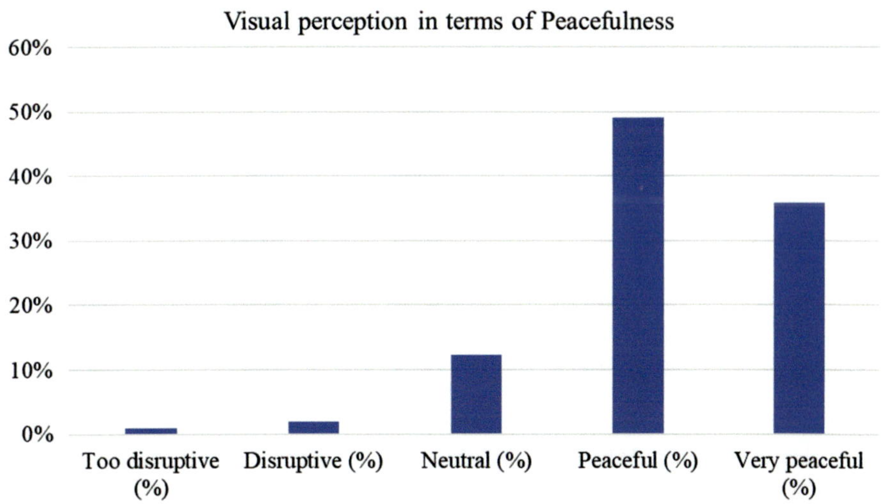

Fig. 13 Visual perception in terms of peacefulness

4.1.5 Importance of Local Cultural Heritage Conservation for Promoting Community Wellbeing

To find out the perceptions of the respondents regarding the importance of cultural heritage conservation for promoting community wellbeing, certain questions were

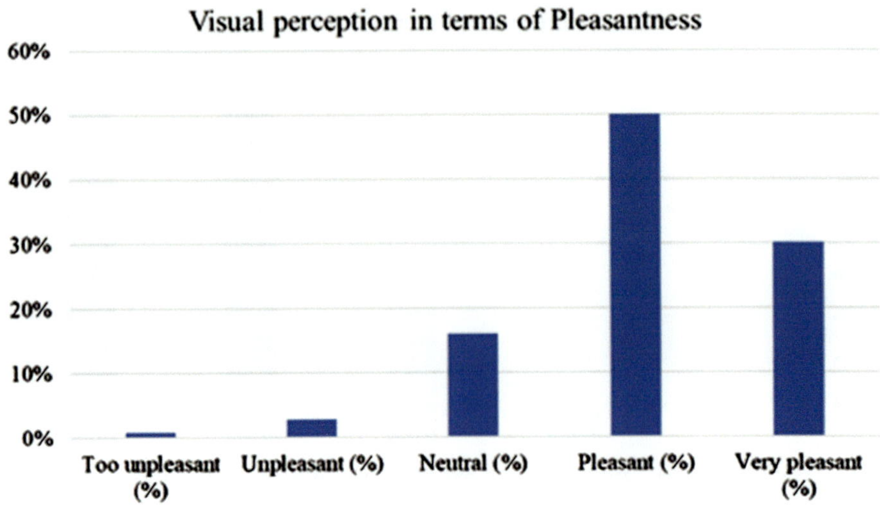

Fig. 14 Visual perception in terms of pleasantness

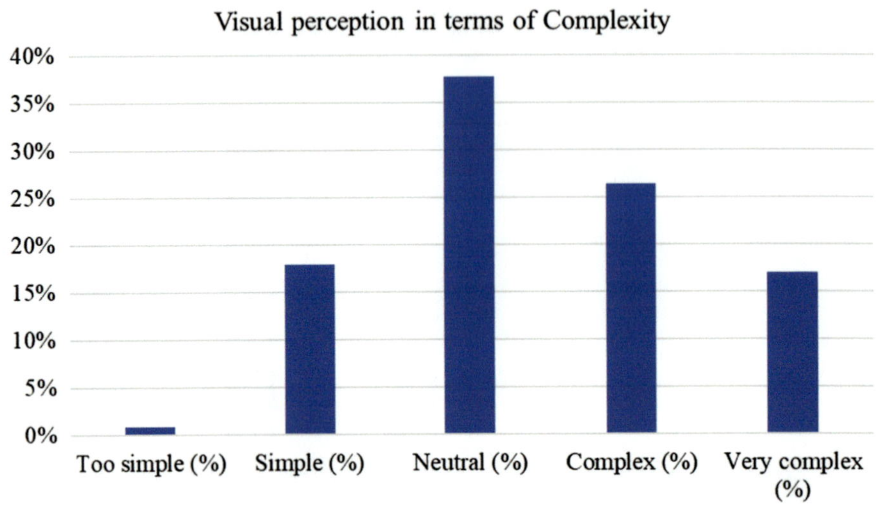

Fig. 15 Visual perception in terms of complexity

asked. The results in Table 7 show that 55% of the respondents feel that it is 'important' to conserve the local cultural heritage for 'maintaining the cultural context and identity of a place,' 45% of the respondents feel that it is 'very important' for the 'process of historical learning and development of knowledge,' 40% feel that it is 'important' for 'promoting local business and job opportunities,' 44% feel that it is 'very important' for 'generating local economy by drawing tourists,' 43% feel it is 'important' for 'reducing the materials and resources consumption,' 36% feel it is

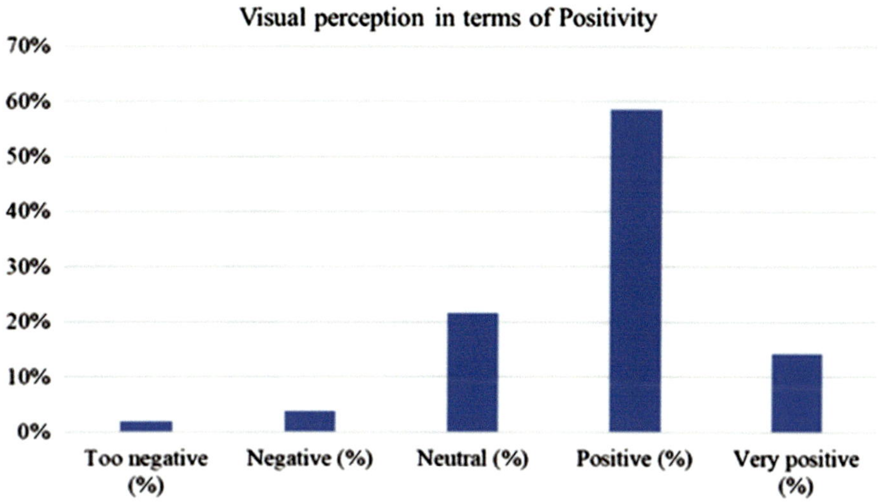

Fig. 16 Visual perception in terms of positivity

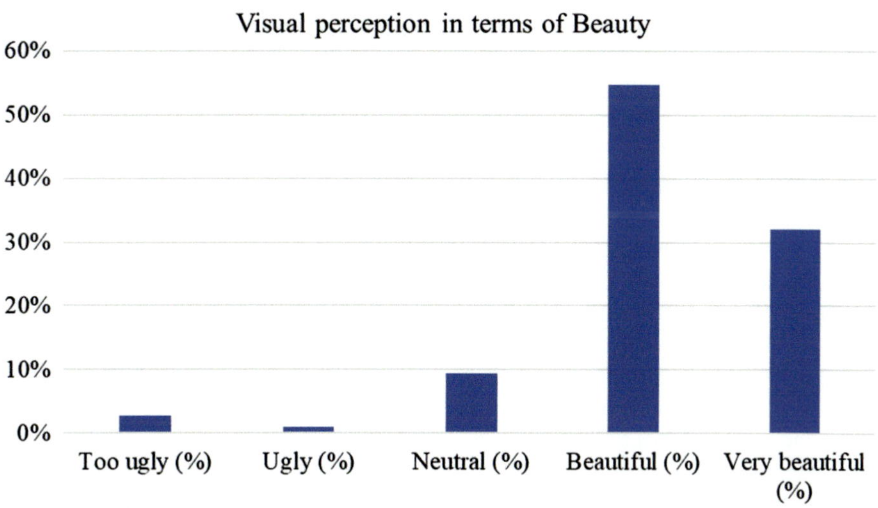

Fig. 17 Visual perception in terms of beauty

'very important' for 'reducing CO_2 emissions' and 45% feel that it is 'important' for 'promoting environmental and ecological awareness.' Out of the seven parameters addressed for community wellbeing, it is observed that 'process of historical learning and development of knowledge' has the highest mean score of 4.2453 whereas, 'reducing the materials and resources consumption' has the lowest mean score of 3.8868. This shows that the respondents feel that conserving the local cultural heritage will help in promoting community wellbeing mainly in terms of historical learning

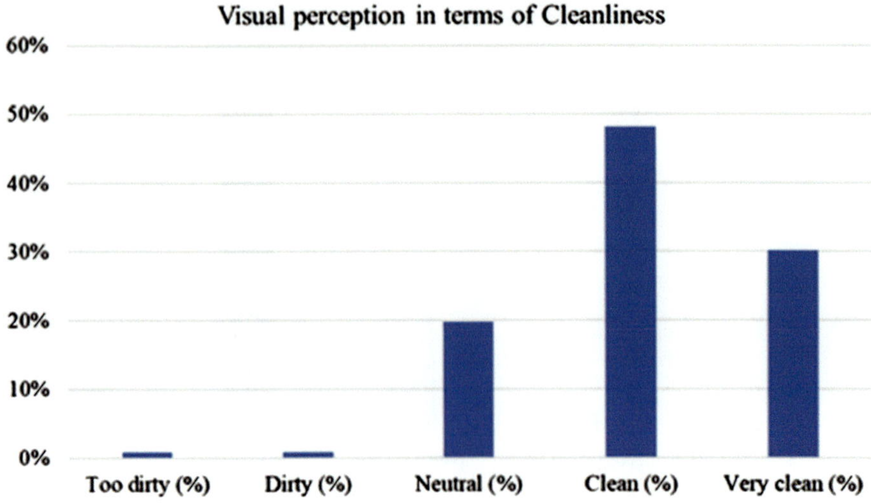

Fig. 18 Visual perception in terms of cleanliness

and development of knowledge. Figure 24 represents the results of the analysis for perception of the respondents regarding the importance of local cultural heritage conservation for promoting community wellbeing.

4.1.6 Participation in the Local Cultural Heritage Conservation

Questions were asked to the respondents to find out their opinions about participating in the process of local cultural heritage conservation. Results in Table 8 show that, most of the respondents (71%) are interested in taking part in the conservation process. 35%, that is most of the respondents want to conserve their local cultural heritage 'for the development of culture.' Most of the respondents (33%) want to participate in the conservation process 'by promoting the local cultural heritage across the world.' Most of the respondents (35%) want 'employment opportunities' as a compensation for participating in the conservation process. Figures 25, 26, 27 and 28 represent the results of the opinion of the respondents regarding participation in the local cultural heritage conservation.

4.1.7 Testing of Correlation Between the Qualitative Aspects with the Age and Gender of the Respondents

Spearman's Rank Correlation has been performed in MS Excel to correlate the qualitative aspects with the age and gender of the respondents as the data is not normally distributed. It has been calculated by using Formula 2. where, n = number of paired observations and

Table 6 Individual wellbeing of the respondents

Parameter	Measure	Percentage (%)	Mean	Standard deviation (SD)
Happiness	Too unhappy	5	4.1226	0.9228
	Unhappy	0		
	Neutral	8		
	Happy	52		
	Very happy	35		
Satisfaction	Too unsatisfied	1	3.9623	0.7798
	Unsatisfied	5		
	Neutral	12		
	Satisfied	61		
	Very satisfied	21		
Pride	Too ashamed	0	4.0000	0.6619
	Ashamed	0		
	Neutral	22		
	Proud	57		
	Very proud	22		
Comfort	Too uncomfortable	0	4.0283	0.6393
	Uncomfortable	1		
	Neutral	16		
	Comfortable	62		
	Very comfortable	21		
Attachment	Too detached	0	4.0755	0.7521
	Detached	2		
	Neutral	19		
	Attached	49		
	Very attached	30		

$$\text{Spearman's 'r'} = 1 - \left(6 \sum d_i^2 / n\left(n^2 - 1\right)\right) \quad (2)$$

d_i^2 = difference between the ranks in the i-th sample.

The null hypotheses (H_0) is that 'There is no significant correlation'.

1a. between the visual perception and age.
1b. between the visual perception and gender.
2a. between the individual wellbeing and age.
2b. between the individual wellbeing and gender.
3a. between the perception regarding the local cultural heritage conservation for promoting community wellbeing and age.

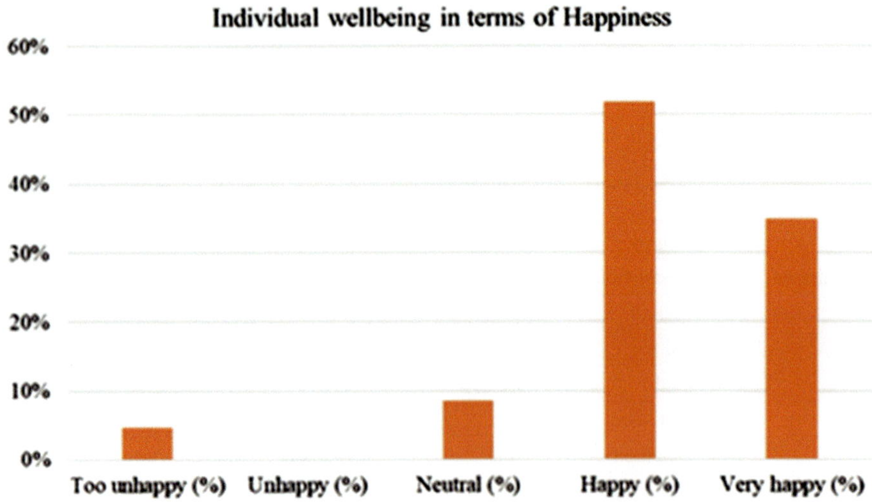

Fig. 19 Individual wellbeing in terms of happiness

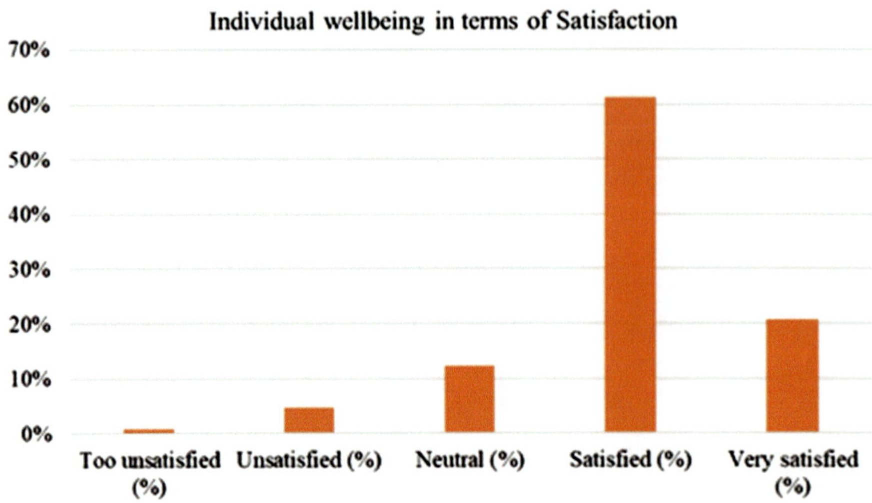

Fig. 20 Individual wellbeing in terms of satisfaction

3b. between the perception regarding the local cultural heritage conservation for promoting community wellbeing and gender.
4a. between the opinions about taking part in the process of local cultural heritage conservation and age.
4b. between the opinion about taking part in the process of local cultural heritage conservation and gender of the respondents in terms of the parameters mentioned in Table 9. The decision rule for accepting or rejecting the hypotheses is to

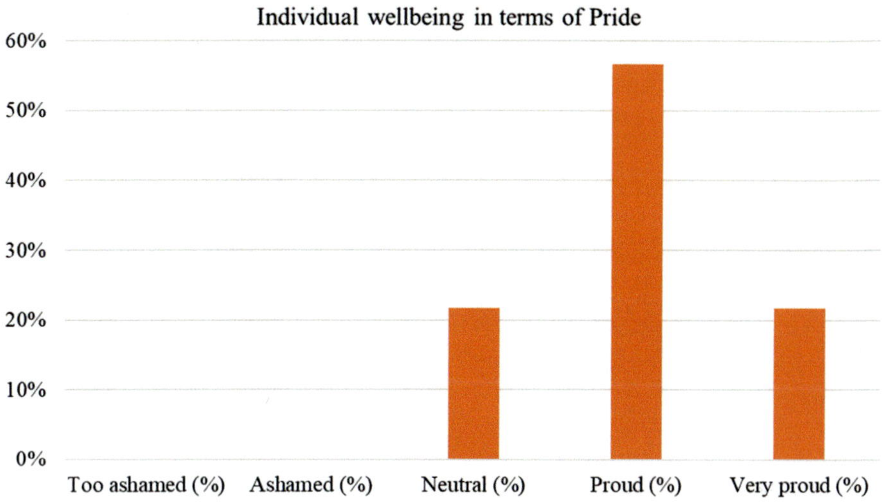

Fig. 21 Individual wellbeing in terms of pride

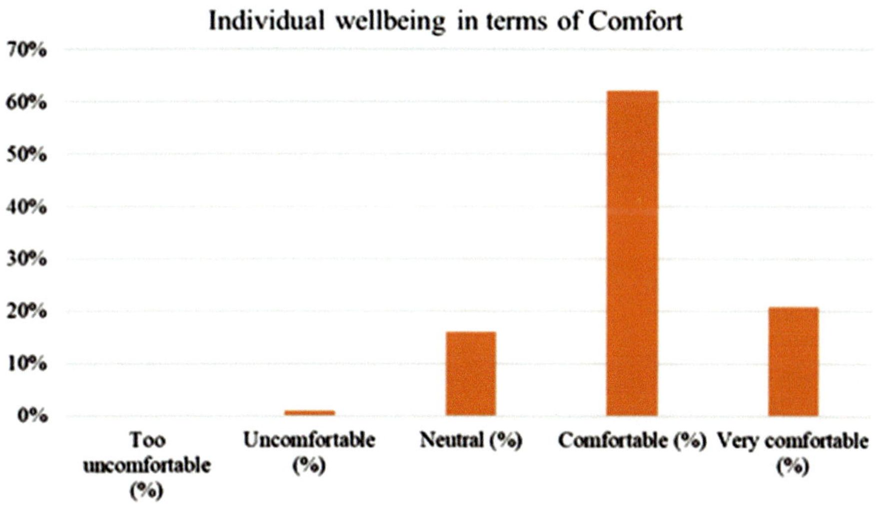

Fig. 22 Individual wellbeing in terms of comfort

compare the level of significance which is 5% that is equal to 0.05 with the p–values obtained from the test.

i. If the p–value is $< = 0.05$, then the null hypothesis (H_0) will be rejected at 5% level of significance and the alternative hypotheses will be accepted.
ii. If the p–value is > 0.05, then the decision will be 'Fail to reject H_0'.

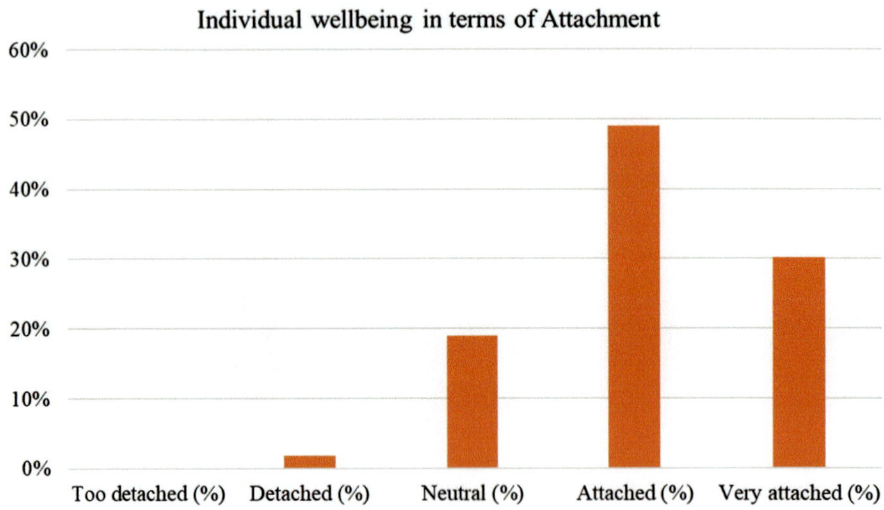

Fig. 23 Individual wellbeing in terms of attachment

Table 7 Importance of local cultural heritage conservation for promoting community wellbeing

Parameters	*TU (%)	U (%)	N (%)	I (%)	VI (%)	Mean	SD
Maintaining the cultural context and identity of a place	3	2	8	55	33	4.1321	0.8516
Process of historical learning and development of knowledge	0	5	11	39	45	4.2453	0.8374
Promoting local business and job opportunities	1	8	15	40	36	4.0094	0.9710
Generating local economy by drawing tourists	1	5	25	25	44	4.0755	0.9827
Reducing the materials and resources consumption	3	6	20	43	28	3.8868	0.9790
Reducing CO_2 emissions	3	6	22	34	36	3.9434	1.0312
Promoting environmental and ecological awareness	2	3	8	45	42	4.2358	0.8569

* TU = Too unimportant, U = Unimportant, N = Neutral, I = Important, VI = Very important

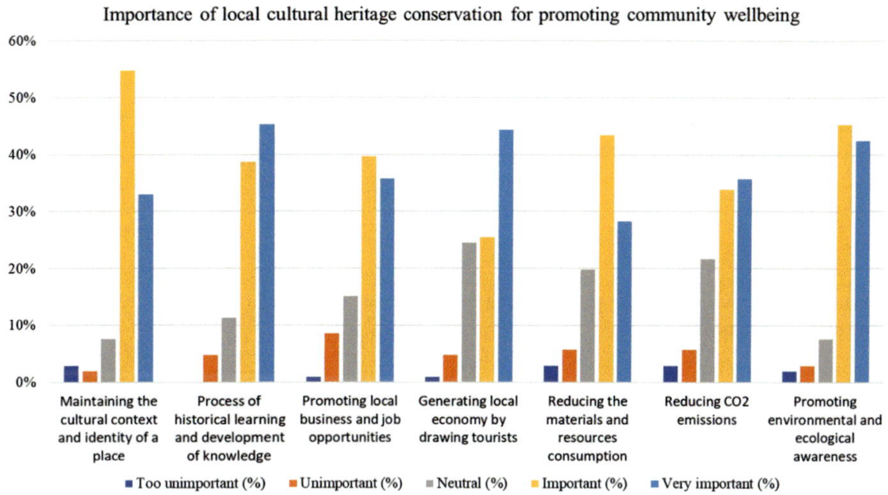

Fig. 24 Results for the importance of local cultural heritage conservation for promoting community wellbeing

Table 9 shows the results of the Spearman's Rank Correlation Test. It is observed that most of the qualitative aspects have a positive correlation with the age and gender of the respondents. However, since, the p–values in most of the cases are greater than 0.05, we 'Fail to reject H_0,' that is the null hypotheses. So, although there is a positive correlation between the qualitative aspects and the age and gender of the respondents, these correlations are not significant. In the case of 'perception regarding local cultural heritage conservation for promoting community wellbeing,' a significant positive correlation is observed between the qualitative aspect in terms of 'reducing the materials and resources consumption' and the gender of the respondents as in this case, the p–value is 0.007 (which is <= 0.05) and the r–value is 0.258.

A significant negative correlation is observed between the age of the respondents and their perception regarding the local cultural heritage conservation for promoting community wellbeing in terms of 'promoting local business and job opportunities', where the p–value is 0.040 and the r–value is −0.200, in terms of 'reducing the materials and resources consumption', where the p–value is 0.006 and r–value is −0.265 and in terms of 'reducing CO_2 emissions', where the p–value is 0.011 and the r–value is −0.246. Also, a significant negative correlation is observed between the age of the respondents and their opinion about taking part in the process of local cultural heritage conservation in terms of 'willingness about participation', where the p–value is 0.012 and the r–value is −0.242 respectively.

Table 8 Participation in the local cultural heritage conservation

Parameter	Measure	Percentage (%)
Willingness to participate in the process of local cultural heritage conservation	Yes	71
	Yes, but it is not my responsibility	4
	Based on livelihood	15
	Not sure	8
	No	3
Reason for your willingness to participate in the conservation process	For the transmission of culture	22
	For the development of culture	35
	For protecting and preserving the cultural identity of the place	22
	For passing on the cultural heritage to the future generation	21
	For gaining associated compensation	1
Way of participation in the conservation process	By controlling the negative impacts on the local heritage	23
	By participating in daily protection and management tasks	8
	By participating in demonstration and awareness meetings	25
	By promoting the local cultural heritage across the world	33
	By supervising the management and development process	10
Expected compensation for participating in the conservation process	Financial assistance	15
	Employment opportunities	35
	Obtain training for tourist reception skills	11
	Revitalized neighbourhoods	10
	No compensation expected	28

4.1.8 Testing of Hypotheses

The hypotheses are tested by using the Mann Whitney U Test in MS Excel as the data is not normally distributed. This is a non-parametric test which serves as an alternative to the independent sample t-test and is used when the assumptions of the t-test are not fulfilled. The decision rule for accepting or rejecting the hypotheses is to compare the level of significance which is 5% that is equal to 0.05 with the p–values obtained from the test.

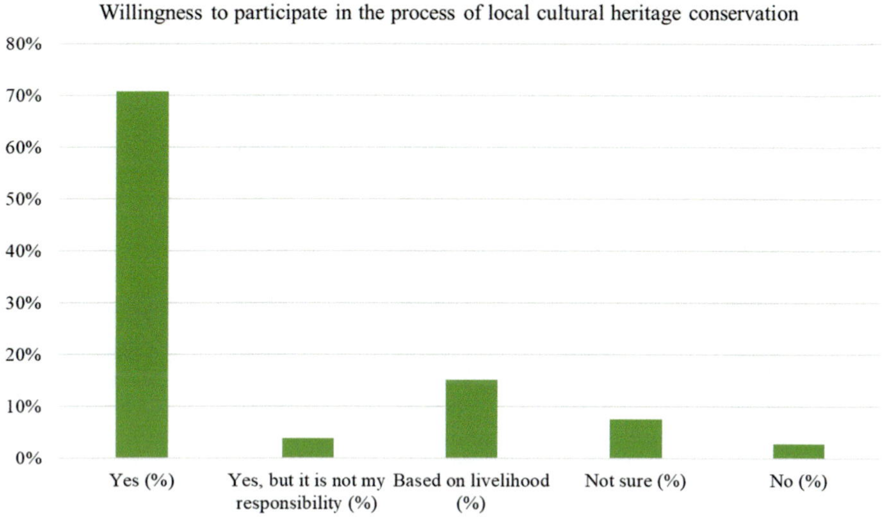

Fig. 25 Willingness to participate in the process of local cultural heritage conservation

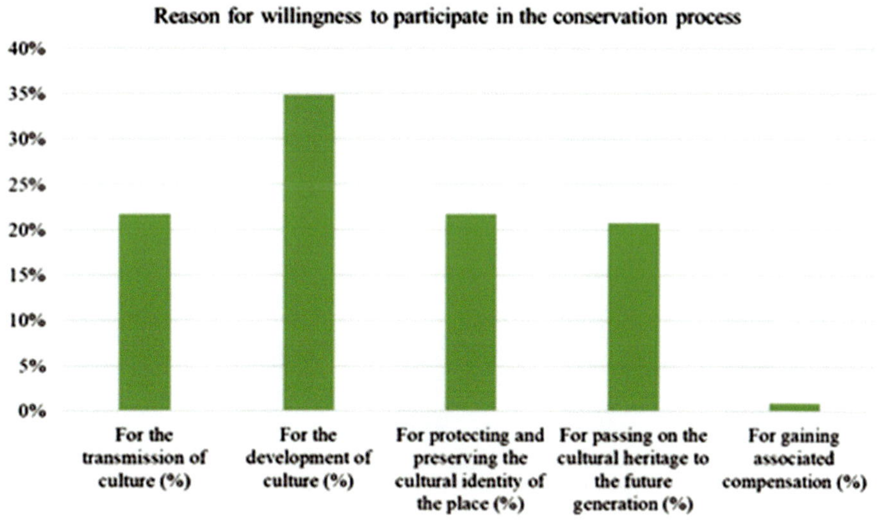

Fig. 26 Reason for willingness to participate in the conservation process

i. If the p–value is <=0.05, then the null hypothesis (H_0) will be rejected at 5% level of significance and the alternative hypotheses will be accepted.
ii. If the p–value is >0.05, then the decision will be 'Fail to reject H_0'.

In the case of this study, the null hypotheses (H_0) is that 'There is no significant difference'.

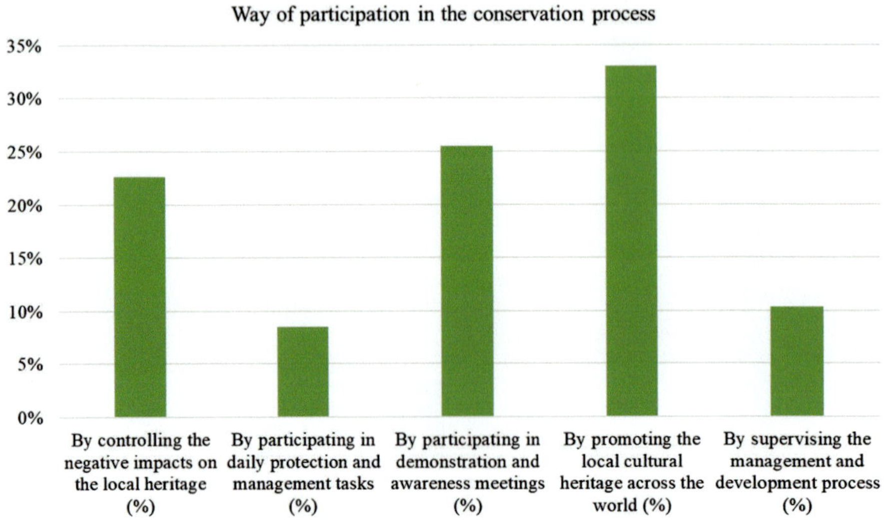

Fig. 27 Way of participation in the conservation process

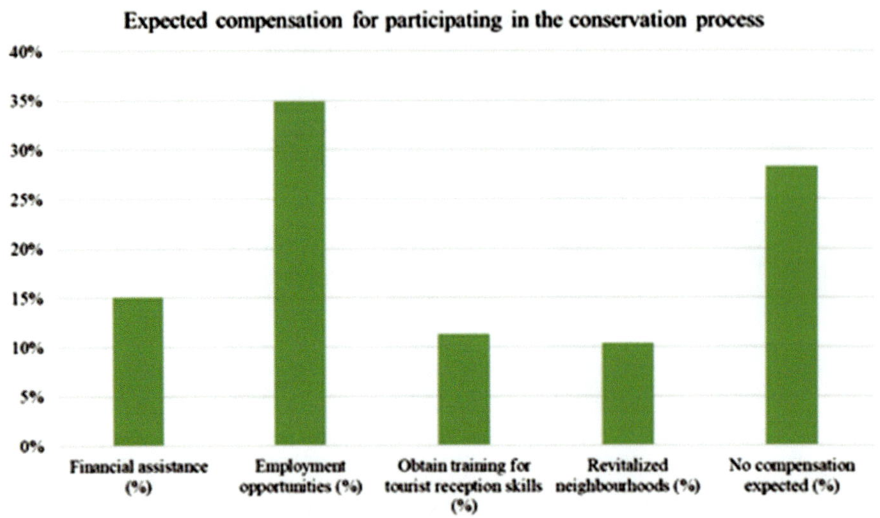

Fig. 28 Expected compensation for participating in the conservation process

i. in the awareness regarding the local cultural heritage,
ii. in the individual wellbeing,
iii. in the perception regarding the local cultural heritage conservation for promoting community wellbeing and
iv. in the opinion about taking part in the process of local cultural heritage conservation among the male and the female teachers in terms of the parameters

Table 9 Results of Spearman's rank correlation test

Null hypotheses (H_0)	Parameters	'r'–values	P–values	Decision
1a: There is no significant correlation between the visual perception and age of the respondents in terms of:	a. Impression	0.183	0.060	Fail to reject H_0
	b. Interesting	0.097	0.323	Fail to reject H_0
	c. Peacefulness	0.076	0.440	Fail to reject H_0
	d. Pleasantness	−0.062	0.527	Fail to reject H_0
	e. Complexity	0.031	0.755	Fail to reject H_0
	f. Positivity	−0.090	0.360	Fail to reject H_0
	g. Beauty	−0.038	0.701	Fail to reject H_0
	h. Cleanliness	−0.063	0.522	Fail to reject H_0
1b: There is no significant correlation between the visual perception and gender of the respondents in terms of:	a. Impression	0.029	0.769	Fail to reject H_0
	b. Interesting	0.095	0.330	Fail to reject H_0
	c. Peacefulness	0.016	0.867	Fail to reject H_0
	d. Pleasantness	0.069	0.479	Fail to reject H_0
	e. Complexity	0.074	0.448	Fail to reject H_0
	f. Positivity	−0.029	0.771	Fail to reject H_0
	g. Beauty	−0.010	0.918	Fail to reject H_0
	h. Cleanliness	−0.012	0.904	Fail to reject H_0
2a: There is no significant correlation between the individual wellbeing and age of the respondents in terms of:	a. Happiness	0.087	0.375	Fail to reject H_0
	b. Satisfaction	0.130	0.183	Fail to reject H_0
	c. Pride	0.012	0.901	Fail to reject H_0
	d. Comfort	0.047	0.630	Fail to reject H_0
	e. Attachment	0.107	0.275	Fail to reject H_0

(continued)

Table 9 (continued)

Null hypotheses (H_0)	Parameters	'r'–values	P–values	Decision
2b: There is no significant correlation between the individual wellbeing and gender of the respondents in terms of:	a. Happiness	0.035	0.725	Fail to reject H_0
	b. Satisfaction	0.000	1.000	Fail to reject H_0
	c. Pride	−0.058	0.554	Fail to reject H_0
	d. Comfort	0.116	0.237	Fail to reject H_0
	e. Attachment	−0.001	0.989	Fail to reject H_0
3a: There is no significant correlation between the perception regarding the local cultural heritage conservation for promoting community wellbeing and the age of the respondents in terms of:	a. Maintaining the cultural context and identity of a place	−0.107	0.276	Fail to reject H_0
	b. Process of historical learning and development of knowledge	−0.028	0.776	Fail to reject H_0
	c. Promoting local business and job opportunities	−0.200	0.040	Rejected
	d. Generating local economy by drawing tourists	−0.153	0.117	Fail to reject H_0
	e. Reducing the materials and resources consumption	−0.265	0.006	Rejected
	f. Reducing CO2 emissions	−0.246	0.011	Rejected
	g. Promoting environmental and ecological awareness	−0.080	0.417	Fail to reject H_0
3b: There is no significant correlation between the perception regarding the local cultural heritage conservation for promoting community wellbeing and gender of the respondents in terms of:	a. Maintaining the cultural context and identity of a place	0.068	0.487	Fail to reject H_0

(continued)

Table 9 (continued)

Null hypotheses (H_0)	Parameters	'r'–values	P–values	Decision
	b. Process of historical learning and development of knowledge	0.147	0.133	Fail to reject H_0
	c. Promoting local business and job opportunities	0.077	0.433	Fail to reject H_0
	d. Generating local economy by drawing tourists	0.025	0.802	Fail to reject H_0
	e. Reducing the materials and resources consumption	0.258	0.007	Rejected
	f. Reducing CO2 emissions	0.023	0.817	Fail to reject H_0
	g. Promoting environmental and ecological awareness	0.053	0.593	Fail to reject H_0
4a: There is no significant correlation between the opinion about taking part in the process of local cultural heritage conservation and the age of the respondents in terms of:	a. Willingness about participation	−0.242	0.012	Rejected
	b. Reason for participation	−0.027	0.780	Fail to reject H_0
	c. Way of participation	0.139	0.155	Fail to reject H_0
	d. Compensation expected for participation	−0.002	0.985	Fail to reject H_0
4b: There is no significant correlation between the opinion about taking part in the process of local cultural heritage conservation and the gender of the respondents in terms of:	a. Willingness about participation	0.087	0.375	Fail to reject H_0
	b. Reason for participation	−0.076	0.441	Fail to reject H_0
	c. Way of participation	0.064	0.511	Fail to reject H_0
	d. Compensation expected for participation	−0.125	0.202	Fail to reject H_0

'r' – values = Spearman's rank correlation values

mentioned in Table 1. The alternative hypotheses have been elaborated in Table 1 with all the parameters.

Table 10 shows the results of the hypotheses testing and whether they have been accepted or rejected. The p–values obtained for almost all the parameters of the four hypotheses are 'greater than 0.05' except for 'reducing the materials and resources consumption' which is 0.013 ($< = 0.05$). So, it can be said that all the alternative hypotheses are being rejected but the one related to 'reducing the materials and resources consumption' is being accepted. So, it can be said that we 'fail to reject H_0' that is the null hypotheses. Therefore, it can be concluded that 'There is no significant difference.'

i. in the awareness regarding the local cultural heritage in terms of all the parameters,
ii. in the individual wellbeing in terms of all the parameters,
iii. in the perception regarding the local cultural heritage conservation for promoting community wellbeing in terms of all the parameters except for 'reducing the materials and resources consumption' and
iv. in the opinion about taking part in the process of local cultural heritage conservation among the male and the female teachers in terms of all the parameters.

4.2 *Discussions*

The results show that most of the respondents are aware of their local cultural heritage. This result aligns with a previous study done by Srivastava (2015). Most of the respondents find their local cultural heritage to be 'peaceful' in terms of visual perception. Most of the respondents feel 'happy' about their local cultural heritage when it comes to their individual wellbeing. Most of the respondents feel that conserving the local cultural heritage will help in promoting community wellbeing mainly in terms of historical learning and development of knowledge. This result aligns with a previous study done by Stolare et al. (2021) where the Swedish primary school teachers supported the fact that visiting the heritage sites contributes to a better historical learning process than simply classroom teaching. Most of the respondents are interested in taking part in the process of local cultural heritage conservation, which is a positive aspect. Most of them want to participate in the conservation process for developing the culture by promoting their local cultural heritage across the world and in return, most of them want employment opportunities as a compensation.

A significant positive correlation is observed between the gender of the respondents and their perception regarding local cultural heritage conservation for promoting community wellbeing in terms of 'reducing the materials and resources consumption.' Whereas, a significant negative correlation is observed between the age of the respondents and their perception regarding the local cultural heritage conservation for promoting community wellbeing in terms of 'promoting local

Table 10 Results of hypotheses testing. (*Source* Author)

Alternative hypotheses	Parameters	P–values	Decision
H1: There is a significant difference in the awareness regarding the local cultural heritage among the male and the female teachers in terms of:	a. Historical structures	0.187	Rejected
	b. Traditional handicrafts	0.542	Rejected
	c. Oral traditions	0.376	Rejected
	d. Performing arts	0.808	Rejected
H2: There is a significant difference in the individual wellbeing of the male and female teachers regarding the local cultural heritage in terms of:	a. Happiness	0.749	Rejected
	b. Satisfaction	1.000	Rejected
	c. Pride	0.595	Rejected
	d. Comfort	0.305	Rejected
	e. Attachment	0.990	Rejected
H3: There is a significant difference in the perception of the male and female teachers regarding the local cultural heritage conservation for promoting community wellbeing in terms of:	a. Maintaining the cultural context and identity of a place	0.532	Rejected
	b. Process of historical learning and development of knowledge	0.166	Rejected
	c. Promoting local business and job opportunities	0.457	Rejected
	d. Generating local economy by drawing tourists	0.812	Rejected
	e. Reducing the materials and resources consumption	0.013	Accepted
	f. Reducing CO2 emissions	0.825	Rejected
	g. Promoting environmental and ecological awareness	0.624	Rejected
H4: There is a significant difference in the opinion of the male and female teachers about taking part in the process of local cultural heritage conservation in terms of:	a. Willingness about participation	0.475	Rejected
	b. Reason for participation	0.455	Rejected
	c. Way of participation	0.523	Rejected
	d. Compensation expected for participation	0.217	Rejected

business and job opportunities,' in terms of 'reducing the materials and resources consumption,' and in terms of 'reducing CO_2 emissions' and between the age of the respondents and their opinion about taking part in the process of local cultural heritage conservation in terms of 'willingness about participation.'

It is found that there is no significant difference in the awareness regarding the local cultural heritage in terms of all the parameters among the male and the female teachers. This result does not align with the previous study that was conducted by Srivastava (2015) where she found that the female teachers are more aware than the male teachers about their cultural heritage. This might be due to the fact that in that study, participants from only a single university were considered as a result of which the findings might not be generalised for all the teachers. In the present study, a more diverse population of teachers have been considered to make the study more rational. Further, it is found that there is no significant difference in the individual wellbeing in terms of all the parameters, in the perception regarding the local cultural heritage conservation for promoting community wellbeing in terms of all the parameters except for 'reducing the materials and resources consumption' and in the opinion about taking part in the process of local cultural heritage conservation among the male and the female teachers in terms of all the parameters.

5 Conclusion

A positive perception and awareness of the local people about their cultural heritage make the conservation efforts a success. In this study, it has been found that the local teachers are quite aware about their cultural heritage and most of them have shown a positive response when asked about their willingness about participating in the local cultural heritage conservation process. Also, it has been found that there is no significant difference in the level of awareness between the male and the female teachers regarding their local cultural heritage. Though most of the respondents are aware, still quite a few of them lack awareness about their oral traditions. So, efforts should be made to strengthen their awareness. Most of the respondents feel that the visits to the heritage sites prove to be more fruitful than simple classroom teaching. Since most of the respondents are willing to safeguard their cultural heritage, it might lead to a positive result when it comes to the conservation of dying arts like the Kondapalli toys, if they receive adequate support from the concerned heritage conservation bodies. This will lead to a participatory approach, where the local people will be involved in the process of protecting and safeguarding their cultural heritage so, that they can maintain the identity and the unique cultural context of their place as well as to pass it on to the future generations. Also, it has been found that most of the respondents want to promote their local cultural heritage across the world. So, this will lead to an increased exposure worldwide, as a result of which a greater number of tourists will be visiting the places, thereby generating a local economy. This will also give rise to increased employment opportunities which most of the respondents are looking for as a compensation for participating in the conservation process. Therefore, all these steps can lead to community wellbeing by creating a community that is sustainable in terms of all the three pillars of sustainability that is environmental, economic and social and also the fourth pillar of sustainability which is the culture.

Though a diverse sample of respondents has been selected for the present study, but as the time period for collecting the data was very short, the number of respondents is also less. Future studies can consider a greater number of respondents to arrive at the conclusions.

References

Ateca-Amestoy V, Villarroya A, Wiesand AJ (2021) Heritage engagement and subjective well-being in the European Union. Sustainability 13(17)

Gallou E (2022) Heritage and pathways to wellbeing: From personal to social benefits, between experience identity and capability shaping. Space and Society, Wellbeing, p 3

Gallou E, Uzzell D, Sofaer J (2022) Perceived place qualities, restorative effects and self-reported wellbeing benefits of visits to heritage sites: Empirical evidence from a visitor survey in England. Space and Society, Wellbeing, p 3

Harsha S, Agarwal S, Kiran CH (2020) Regional flood forecasting using SWMM for urban catchment. Int J Eng Adv Technol (IJEAT) 9(3):1027–1031

Innes A, Scholar HF, Haragalova J, Sharma M (2021) 'You come because it is an interesting place': the impact of attending a heritage programme on the well-being of people living with dementia and their care partners. Dementia 20(6):2133–2151

Kee T, Chau KW (2019) Economic sustainability of heritage conservation in Hong Kong: the impact of heritage buildings on adjacent property prices. Sustain Dev 28(1):308–319

Lerario A (2022) The role of built heritage for sustainable development goals: from statement to action. Heritage 5(3):2444–2463

Sektani HHJ, Khayat M, Mohammadi M, Roders AP (2021) Erbil city built heritage and wellbeing: An assessment of local perceptions using the semantic differential scale. Sustainability 13(7)

Shankar B, Swamy C (2013) Creating awareness for heritage conservation in the City of Mysore: issues and policies. International Journal of Modern Engineering Research (IJMER) 3(2):698–703

Shimray SR, Ramaiah CK (2019) Cultural heritage awareness among students of Pondicherry University: a study. Libr Philos Pract (e-journal)

Srivastava S (2015) A study of awareness of cultural heritage among the teachers at University level. Univ J Educ Res 3(5):336–344

Stolare M, Ludvigsson D, Trenter C (2021) The educational power of heritage sites. Hist Educ Res J 18(2):264–279

Taçon PS, Baker S (2019) New and emerging challenges to heritage and well-being: a critical review. Heritage 2(2):1300–1315

Wiseman J, Brasher K (2008) Community wellbeing in an unwell world: trends, challenges, and possibilities. J Public Health Policy 29(3), 353–366

Evaluating the Adaptive Reuse Potential of Chettinad Houses for Heritage Conservation

Palaniappan Alamu Priya

Abstract The Chettinad region is known for its rich residential architecture. The grandeur mansions bear testimony to the successful banking and trading community of the Nattukottai Nagarathars. However, in recent times, a good lot of buildings have become dilapidated due to the migration of residents to other places in search of better opportunities. This paper aims to analyze the adaptive reuse potential of the region to conserve the palatial houses for the development of culture, heritage, and way of life for future generations. It involves a brief historical research on the evolution and principles of the community and a combination of models such as the ARP model and the Delphi method to analyze its potential. After due consideration of the physical obsolescence, economic obsolescence, functional obsolescence, technological obsolescence, social obsolescence, and legal obsolescence, the ARP was found to be 31.3, which shows that the building has a moderate value for adaptive reuse. Hence, adaptive reuse is an apt choice to sustain the life of the building in this context. Besides, the Delphi Model that is based on people's perceptions also showed the need to strategize the adaptive reuse of Chettinad houses.

Keywords Adaptive reuse · ARP model · Chettinad region · Delphi · Dilapidated · SDG11

Abbreviation

ARP	Adaptive Reuse Potential
UNESCO	United Nations Educational, Scientific and Cultural Organization
WHS	World Heritage Sites

P. A. Priya (✉)
School of Planning and Architecture, Vijayawada, India
e-mail: alamupriya16@gmail.com

Vijayalaxmi J. (ed.), *Conservation of Built Heritage in India*, Advances in 21st Century Human Settlements, https://doi.org/10.1007/978-981-99-8335-3_9

1 Introduction

The architecture of the Chettinad region, an indigenous amalgamation of vernacular architecture and western influences, has always been of great recognition. The houses are being abandoned in recent times.

Studies show that UNESCO has prepared a draft to conserve the palatial houses involving multidisciplinary approaches (UNESCO 2014). They are justified to be of significance since they possess outstanding physical characteristics of urban and rural planning. The climate and the surroundings were always taken into consideration. They are also known for their unique lifecycle rituals and beliefs. They are important sites promoting tourism and thus have chances to even increase the economic viability of the place (UNESCO 2010; Meenakshi 2010).

1.1 Importance of Conservation of Architectural Heritage of Chettinad Region

The main town of the region, Karaikudi is located 90 km from other major World Heritage Sites (WHS) like Thanjavur, Trichy, and Madurai. This region also displays some of the most interesting water management systems and drainage patterns. They are also known for their unique lifecycle rituals and beliefs. They are important sites promoting tourism and thus have chances to even increase the economic viability of the place (UNESCO 2010; Kiruthiga and Thirumaran 2017). The target 11.4 of SDG11 also suggests the safeguarding of cultural and natural heritage as an important aspect of sustainable development.

1.2 Values in the Built Heritage of the Chettinad Region

People inherit a physical environment that is very often associated with their identity in some way along the social process towards the heritage of a region (Australia ICOMOS 1979; Riegl 1996). Earlier, the heritage sites were narrated with their prehistoric and cultural significance (Hearn 1990; Ruskin 1849). The heritage value of a region relies on the respective cultural, intellectual, historical, and psychological credentials of a region (Australia ICOMOS 1979; Lipe 1983). Tangible and intangible heritages are the two attributes of built heritage, including all the aspects of buildings that contribute to the spatial configuration of the town with regard to size, color, materials, and detailed decorations. The attributes describe the identity of a region. This research exhibits the importance of the contribution of values to the built heritage. The different values attributed to the Chettinad region include various indicators and their variables to justify their importance. They include:

- Architectural—Construction techniques, materials, aesthetics, styles, etc.
- Socio-cultural—Age, historical significance, culture, social, etc.
- Environmental—Landscape setting, water bodies, ooranis, ecological balance, etc.
- Economic—Use value, non-use value, option have, and bequest.

Studies show that these values have to be addressed to preserve them for future generations or they will become of disregard (Rajivkumar 2019).

1.3 Resident's Attitude Towards Preservation of Chettinad Homes

The future of the Chettinad houses stands uncertain as a lot of residents have left their villages and houses in search of better opportunities. Not all houses are of the same fate, but some of the houses export their architectural components to western countries as antiquities before they get demolished. Thus, the issues affecting the region include economic conditions and land speculation, massive desertion, the antiquities market, lack of protection and regulation in a vast territory, and lack of awareness and public participation. Thus, surveys were conducted by preparing questionnaires that asked the residents how they felt about the preservation of their houses, how they valued the built heritage, and what they thought were the existing barriers. Results stated that the residents wanted to preserve them as they hold memories of the place, but could not do so because of economic issues, the land holding more value than the house itself, and few houses were owned by multiple people (Rajivkumar 2019).

1.4 Adaptive Reuse, Restoration of a Chettinad Mansion

This paper states that it is necessary to reuse a building to ensure its economic, social, and cultural stability as witnessed in the adaptive reuse of a 110-year-old mansion, Chidambara Vilas, which is turned into a heritage hotel. This paper talks about the different stages of adaptive reuse—firstly the planning to understand how new functions have been incorporated, secondly the material restoration, and later the addition of new services.

Conclusions state that the ethics of conservation have been kept in mind during adaptive reuse. The new functions adapt well to the historic fabric and are thus well-integrated (Dhakshinamoorthi et al. 2014).

1.5 Adaptive Reuse Potential of Buildings in Heritage Areas

Adaptive reuse becomes very essential for sustainability. The decision-making process is often a difficult stage to analyze if the heritage structure requires reuse. Thus, a model has been established called the Delphi method with combined views from different professionals on the degree of various factors that affect the building's potential. This also enhances the better allocation of resources. Interviews have also been conducted and then applied to the Grand Canal Area, China to generate the potential and priority of reuse. Findings show that this is an effective process to engage the public in decision-making catering to variables like facilities, landscape quality, functions, structural stability, etc. (Chen and Hong 2017).

1.6 Building Adaptive Reuse Opportunities in Hong Kong

The paper talks about an adaptive reuse model that aims to adaptively reuse the Hong Kong market area. The model could enhance the decision-making process towards more sustainable practices and rank the buildings that have a high potential for adaptive reuse. The research findings are based on two ideas: improving existing building adaptation results in better use of locally available resources and lower demand on the environment. Additionally, national heritage is protected. The model considers the building's expected physical life as well as its current age, both in years. Physical, functional, economic, social, technological, and legal obsolescence must also be evaluated. The ARP values are calculated, and if they are greater than 50, they are thought to have a significant potential for reuse. Values 20–49 indicate moderate potential, while values 1–19 indicate low potential. There is no potential with an ARP value of zero.

1.7 Adaptive Reuse Analysis Through Multi-Criteria Decision-Making

Decisions involving the tangible and intangible components of numerous interest groups were employed to assess the historic reuse of the Sun Yat-Sen Historical Museum. The priority of reuse alternatives was analyzed through a combination of the Delphi method and an analytic process. The criteria had to be developed and then the evaluation process. The opinions of a list of experts from various fields have then been aggregated. The results revealed that most of the sustainable reuse options were for community use, followed by commercial, education and exhibition, and composite activities (Chen et al. 2018).

2 Study Area

2.1 Chettinad Region

The Chettinad region is situated in the Sivagangai district. The term 'Chettinadu' means 'land of the Chettiars'. It is located in Tamil Nadu, India (Fig. 1).

2.1.1 History

The Chettinad region has a very long past dating back to the thirteenth century. They began as gem traders in a town called Santhiapuri, Naganadu, after which they had to migrate to Kanchipuram due to the torture of the King. They further migrated to Poompuhar in the Chola kingdom due to unjustified taxes and levies imposed. Here, they became prime traders in salt and received the honor of crowning Kings. The Port at Poompuhar was submerged due to a Tsunami. It was at that time the Pandiya Kind invited the Chettiars to enrich his kingdom which was suffering from war destruction. He thus gifted an arid region to the Chettiars which is the current Chettinad region. With the arrival of the British in India, they grew into money lenders and started involving in merchant banking. Maritime trade flourished and they invested all their wealth in the construction of their palatial mansions (Dokras 2021). During the post-independence period, they faced bankruptcy, as a result of which they had to migrate to other places in search of new jobs. That becomes a major reason why the houses remain abandoned (Subramanian 2018).

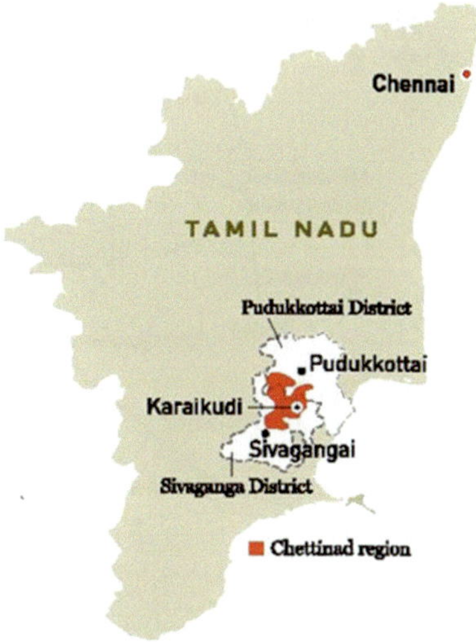

Fig. 1 Location of Chettinad region 2018 (*Source* Subramanian)

2.1.2 Demographics

In the nineteenth and early twentieth centuries, many residents of Chettinad were trading in South and Southeast Asia, by 2010, out of the original 96 villages, only 74 villages remained. They were organized in clusters and spread over 1,550 square kilometers in the Districts of Pudukottai and Sivagangai (Fig. 2) in Tamil Nadu (Rajivkumar 2019).

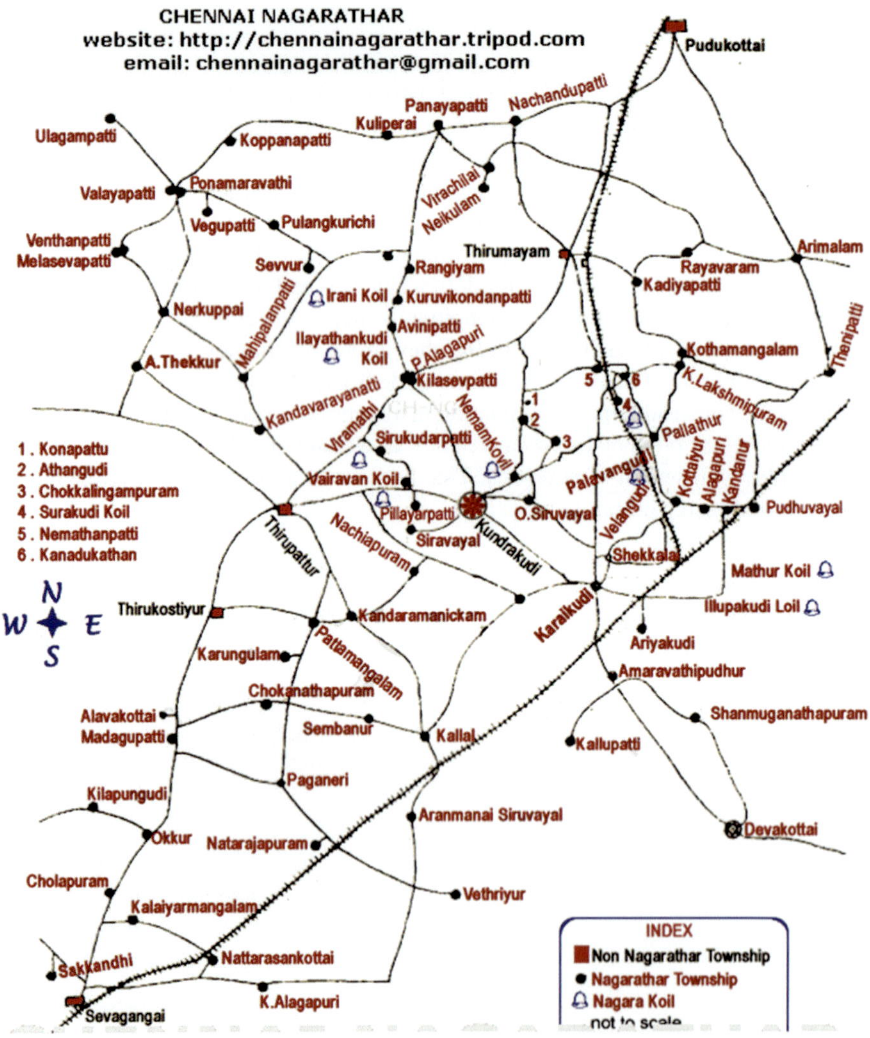

Fig. 2 Map of Chettinad region (*Source* Chennai Nagarathar, n.d.)

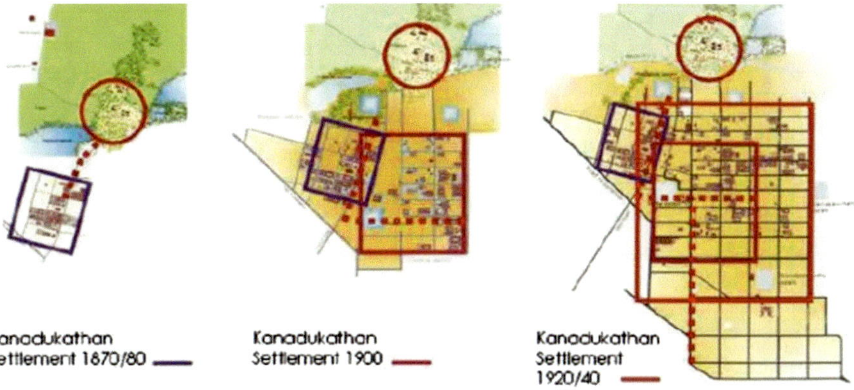

Fig. 3 Example of grid-iron town planning in Kanadukathan, a Chettinad settlement 2015 (*Source* Natarajan)

2.1.3 Town Planning

Town planning and house constructions are contained in rectangle plots along one longitudinal axis to orthogonal street layouts as shown in Fig. 3. The houses' front doors were towards one street and their rear doors were towards a parallel running street. Since those days, their grid system of villages has additionally implemented sophisticated water harvesting systems.

2.1.4 Architecture

Before venturing into the general planning of the region, the historical events mentioned above had major impacts on their planning principles stated below (UNESCO 2014):

- The Chettiars always migrated as groups and this practice had a profound effect on their spatial planning and division of spaces in the houses. They lived as joint families and mostly all spaces in the house were commonly used by everyone.
- Due to the Tsunami that occurred, the community always had a fear of the sea, and this was reflected in the elevated plinths of the house which also depicted grandeur, wealth, and status.
- New construction materials and techniques were discovered to beat the summer heat as they settled in an arid region.
- Trade links had an important influence on the architecture of the Chettinad region. They imported a lot of materials like wood, satinwood, glass, marble, ceiling tiles, and chandeliers from Burma, Ceylon, Belgium, Italy, England, and Morocco respectively.

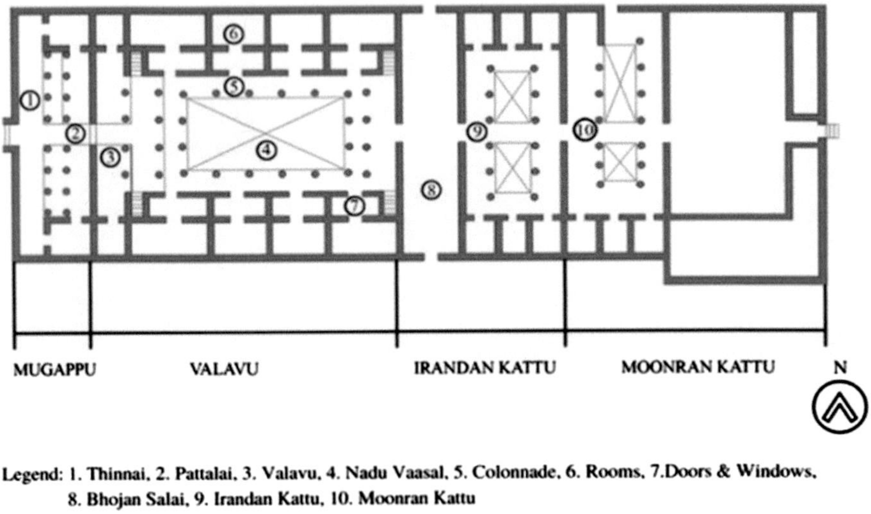

Fig. 4 Plan of a typical Chettinad house 2015 (*Source* Michell and Ramamrutham)

- Culture and tradition also played an important role in shaping the architectural planning of the mansions. Social order is of great importance when it comes to the allocation of spaces.

The early houses were bilaterally symmetrical, mostly single-storied as shown in Fig. 4. There is a gradual increase in the aspects of the usage of the house from public to private as we progress. The spaces of a typical Chettinad house include (Michell and Ramamrutham 2015; Patwardhan 2018):

1. **Mugappu—**It is the reception or the entrance of the house, a public space abutting the street.
2. **Thinnai—**This outer public space has raised platforms on either side of the central corridor, where the host would entertain guests. It is like a verandah space.
3. **Muttram/Nadu Vaasal—**This is the interior, central courtyard space that is used for different ceremonies starting from birth to death.
4. **Valavu—**This is the living area of the house. They consist of the Irattai Veedu—the rooms. Each of the small rooms is the property of one married son. They are usually two rooms connected by a single door.
5. **Nadai—**The pillared walking space around the courtyard connecting to the Irandaam and Moonam kattu is the nadaipathai.
6. **Iradan kattu—**Used for dining purposes during family occasions. It has storage rooms for storing/kitchenware.
7. **Moonam kattu—**The rear hall served as the woman's domain which consisted of the kitchen and a space for other domestic activities.

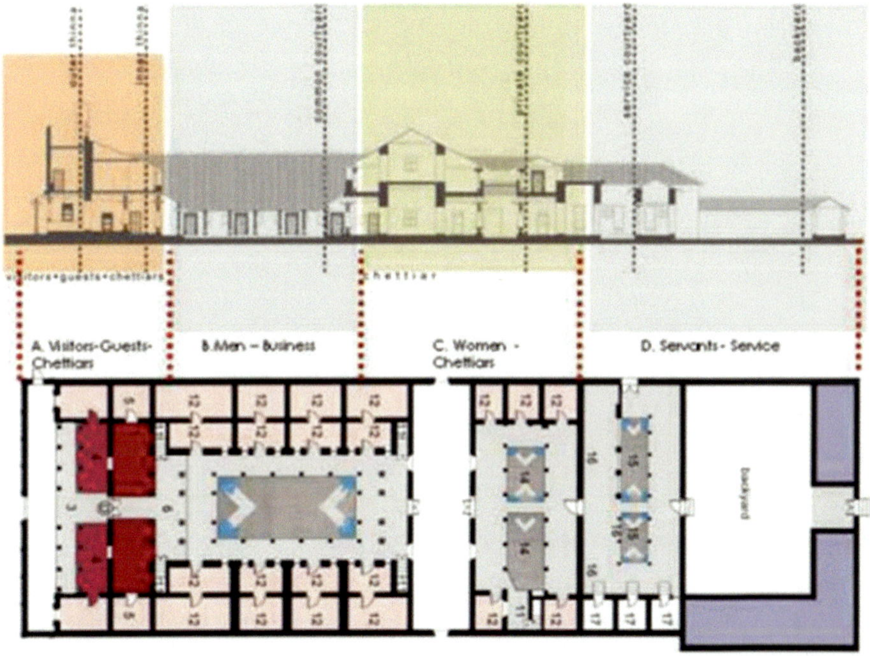

Fig. 3 Spatial planning and hierarchy 2015 (*Source* Natarajan)

Fig. 4 Chettinad mansion elevation (*Source* Behance n.d.)

8. **Thotam**—This is the garden space at the rear of the house since earlier households used this space to rear cattle.

During the period 1850–1950, the houses in the Chettinad region underwent a transformation, in built form and the materials used, from simple dwellings to large independent mansions. The bilateral symmetry concepts were slightly altered by

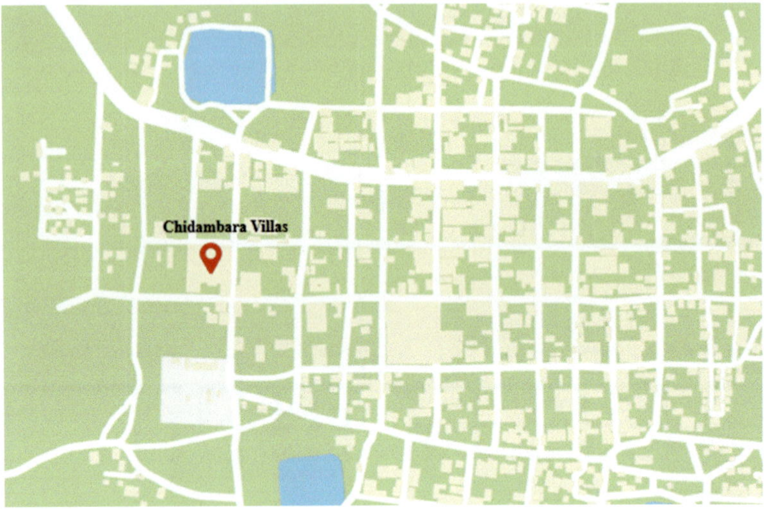

Fig. 5 Location of Chidambara villas (*Source* Author)

the addition of spaces like Bhojan Salai. As the traders grew wealthy, they began constructing double-storied structures.

The house was also planned according to spatial usage (Fig. 5). The thinnai portions were spaces for the public that acted as gathering places. The courtyards were the spaces for the men, followed by the irandaam kattu for domestic activities, and the moonam kattu and the thottam for the servants of the house.

The elevations of the houses, as shown in Fig. 6, were a combination of the Art Deco, Victorian, and Art Noveau styles. The Art Noveau style can be evident from the arched windows, extensive use of curved forms, and use of stained glass and balustrades. The double columns in the double-storied structures were elements of Victorian architecture. The triangular and zig-zag patterns referred to the Art Deco styles (Patwardhan 2018).

3 Materials and Methods

The significance of this research aims at establishing an evaluation tool to estimate the useful life of a building based on obsolescence criteria. Once established, it can be used to further compare the current age of the building and the expected age of the structure to evaluate the adaptive reuse potential.

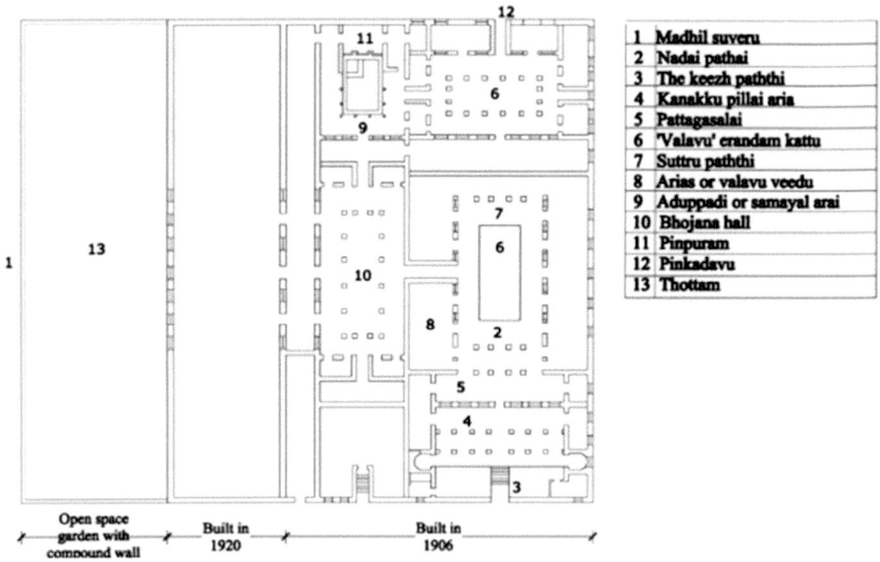

Fig. 6 Ground floor plan—original mansion 2019 (*Source* Mohan)

3.1 Case Study—Chidambara Villas

To analyze the potential of the Chettinad mansions, Chidambara Villas—a 110-year-old heritage Chettiar house that is adaptively reused into a heritage hotel is chosen as a case study. The data are collected from secondary sources (Rajmohan 2016).

The Chettinad heritage hotel is situated in Ramachandrapuram, Kadiapatti. Chidambara Vilas' history began in 1884 when members of an influential Chettiar family left for abroad. The mansion's construction began on a one-acre plot (Fig. 7) donated by Maharaja of Pudukottai. Over 110 years ago, it took 7 years (1900–1907) to complete the house for a sum of 7 lakhs. In 1920, an addition was built.

The mansion not being in use in the later years was given on lease to the Sangam Group (30 years) for reuse and restoration of the house into a heritage resort. The process took 3 years, which was nearly half the time it took to construct the actual home.

3.2 Conservation Theories

The following conservation theories and practices were adopted during the restoration.

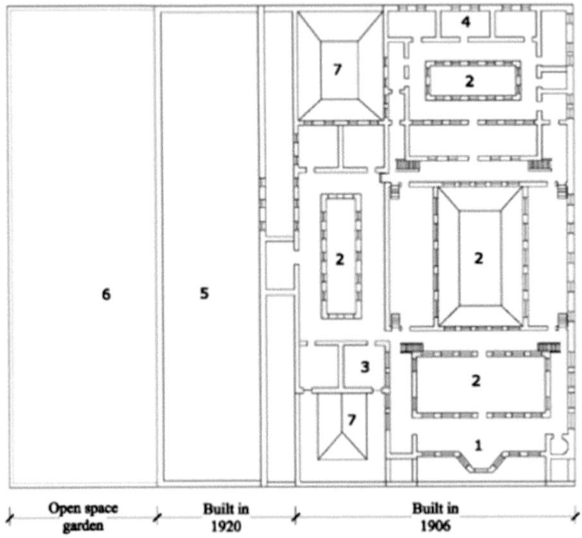

S.No	SPACES
1	Arai
2	Double storeyed hall
3	U I arai
4	Arai
5	Open terrace
6	Thottam
7	Slope roof

Fig. 7 First floor plan—original mansion 2019 (*Source* Mohan)

3.2.1 Adaptive Reuse

The process of reusing an existing building for an intent other than the one for which it was initially designed/built is referred to as adaptive reuse. It is also referred to as conversion or recycling. Before rejuvenation and reconstruction of a structure, the first step would be to ensure that the completed product will meet the market demand, be completely useful for its new purpose, and be competitively priced. A longer time gap between the two would favor adaptive reuse strategies.

The research outcomes would establish the fact that adaptive reuse will lead to more effective use of the building, more use of local materials, less demand on the environment, and a good economic generator. On another note, heritage is also conserved and preserved. This was done in two ways, either assigning similar usage or with minor interventions in design like closing the doorways, the addition of a partition wall, and the provision of lighting fixtures. The original plans of the mansions are depicted in Figs. 8 and 9.

3.2.2 Restoration Process

The restoration process is done either by regular restoration measures or material restoration measures.

Fig. 8 Before and after—color schemes 2019 (*Source* Mohan)

Fig. 9 Cultural artifacts 2019 (*Source* Mohan)

Restoration Measures Include

- Cleaning of the premises
- Cleaning the architectural elements like arches, floors, façade elements, etc.
- Granite tiles replacement in the first courtyard from a local quarry.
- Painting in accordance with the Chettinad color schemes (Fig. 10).
- Covering open wells using iron rods to preserve them as a cultural artifact (Fig. 11).

Material Restoring Measures Include

- Replacing the damaged Athangudi tiles on the floor as well as the roof (Fig. 12), while restoring the existing patterns and colors.

Fig. 10 Preparation of Athangudi tiles 2019 (*Source* Mohan)

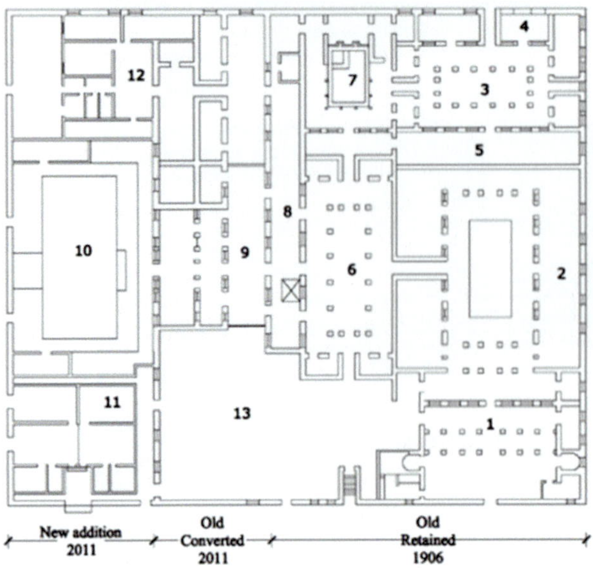

S.NO	SPACES
1	Reception office
2	Store room
3	Experimental kitchen
4	Rooms
5	Dinning hall
6	Restaurant
7	Modern kitchen
8	Service space
9	Recreation
10	Swimming pool
11	Ancillar facilities
12	Services
13	Owners residence

Fig. 11 Ground floor plan—heritage hotel 2019 (*Source* Mohan)

- Chettinad plaster prepared on-site by the local craftsmen and applied to the walls.

3.2.3 Addition of Spaces and Modern Services

The existing mansion was extended to accommodate modern hotel requirements such as guest rooms, a swimming pool, services such as a lift, lighting, air-conditioning, water supply, water recycling, drainage, water recycling, and a treatment plant. The new spaces are in continuity with the existing spaces (Figs. 13 and 14).

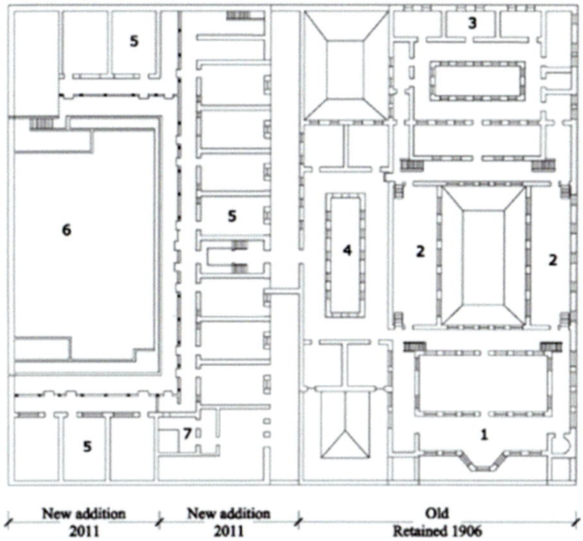

S.No	SPACES
1	Exhibition
2	Store
3	Store
4	Royal hall
5	Guest room
6	Swimming pool
7	House keeping

Fig. 12 First floor plan—heritage hotel 2019 (*Source* Mohan)

Fig. 13 Guest rooms 2019 (*Source* Mohan)

Fig. 14 Detailing of furniture and electrical accessories 2019 (*Source* Mohan)

Fig. 15 Lighting fixtures 2019 (*Source* Mohan)

The new constructions are also done following the Chettinad style (Fig. 15):

- Providing variations of thinnai spaces within the guest rooms as a sill extension.
- Usage of typical Chettinad furniture (Fig. 16).
- Using materials for the extensions as per the old mansion such as baked bricks, lime wash, and Chettinad plaster. Even old doors and frames are bought from dilapidated houses.
- Additional spaces like swimming pools are built with sloped roofs using terracotta tiles.
- The interiors of the guest elevators also carry details of the Chettinad railing done using teak wood.

Thus, the adaptive reuse process has not just changed the interior layouts of the building and made it functionally available for the people but has also aesthetically transformed the facades (Fig. 18) and interiors of the mansion, being careful not to tamper with the heritage at any given point in time.

Fig. 16 Design of extensions 2019 (*Source* Mohan)

3.3 ARP Model

This paper discusses the conceptual framework for identifying and ranking the ARP for the present buildings. It is applicable to structures in all countries, but it is specifically applied to the Chettinad region. The method includes the building's current age as well as its estimated life. Physical, functional, economic, social, technological, and legal obsolescence must also be assessed. The values of reuse potential obtained are denoted in the form of percentages.

- **Physical obsolescence** is determined by the examination of maintenance policies and performances. According to the developed scale, buildings with a high maintenance budget are given a 0% reduction, and those with a low-maintenance budget receive a 20% reduction. Interim scores of 10% reduction can also be given to buildings with normal budgets.
- **Economic obsolescence** is measured by the location of a building in a city center or central business district. Structures sited amidst high population densities are given a 0% reduction as per scale, and buildings amidst low population densities are evaluated as a 20% reduction. Interim scores of 10% reduction for average population density.
- **Functional obsolescence** is determined by the flexibility of the design of the building. The useful life of a building is reduced if the building layouts are not flexible to change. 0% reduction for buildings where customers stop using the

services after a long time, and 20% reduction where service usage stops shortly. Interim values of 10% reduction for average usage durations.

- **Technological obsolescence** is determined by the building's operational energy use. If high levels of energy are used by the building to provide occupant comfort, then the building's useful life might reduce. Buildings with lesser energy demands are given a 0% reduction, high energy demands receive a 20% reduction, and an interim score of a 10% reduction for conventional energy performances.
- **Social obsolescence** could be measured through the relationship of the building function with the marketplace. If the building's feasibility is based on external income, then the building's useful life reduces. Buildings with fully owned and occupied space receive a 0% reduction, fully rented spaces receive a 20% reduction, and balanced rent and ownership get an interim value of a 10% reduction.
- **Legal obsolescence** can be determined by the quality of the original design. If designed to low standards, then the useful life reduces. The scale denoted a 0% reduction for high-quality designed buildings, a 20% reduction for lower-quality designs, and a midway score of 10% for average quality.

$$\text{Useful life } (L_u) = \frac{L_p}{\left(1 + \sum_{i=1}^{6} O_i\right)^{L_p}},$$

where L_p denotes physical life in years; O1—physical obsolescence; O2—economic obsolescence; O3—functional obsolescence; O4—technical obsolescence; O5—social obsolescence, and O6 denotes legal obsolescence.

EL_u (Effective useful life), EL_b (effective building age), and EL_p (effective physical life) respectively, are calculated by multiplying L_u, L_b, and L_p with 100 and dividing by L_p.

The values of increasing and decreasing ARP are then calculated using the formulas:

$$\text{ARP}_{\text{(increasing)}} = 100 - \left(\frac{\text{EL}_u^2 / 100}{\text{EL}_u}\right) \times \text{EL}_b,$$

$$\text{ARP}_{\text{(decreasing)}} = 100 - \frac{(\text{EL}_u^2 / 100)}{100 - \text{EL}_u} \times (100 - \text{EL}_b),$$

The value of ARP increasing is used if the effective building age is lower than or equal to the useful life.

$$(\text{EL}_b \leq \text{EL}_u)$$

The value of ARP decreasing is used if the effective building age is greater than equal to the useful life.

$$(EL_b \geq EL_u)$$

3.4 Delphi Method

This evaluation model takes into consideration the views of different people including professionals and tourists who are most affected by the adaptive reuse projects, regarding the aspects and variables that affect a building asset's adaptability. It is a structured and effective communication process that deals with complex issues of adaptive reuse and other fields.

A secondary survey is circulated amongst the various people either directly or indirectly related to the project. Each variable is evaluated on a scale (low to high) and then the overall scores are analyzed. The building can be ranked for its adaptive reuse potential based on the overall scores. Further, the positive and negative variables for the adaptive reuse potential are identified so that the implementation of specific strategies and adaptations in the designs can be done.

The parameters based on which the survey is conducted are listed below:

- **Life Cycle**
  1. Structure
  2. Function
- **Quality**
  1. Style
  2. Interiors
  3. Facility
  4. Material
- **Location**
  1. Surroundings
  2. Site conditions
- **Transport**
  1. Pedestrian
  2. Public transport
  3. Road systems
- **Space**
  1. Green space
  2. External spatial legibility

- **Economic Value**
  1. Economic return
  2. Use value
- **Cultural Value**
  1. Cultural beliefs
  2. Historical value
- **Social Value**
  1. Harmonious with neighboring buildings
  2. Facilitates social interactions
  3. Public education
  4. Value exhibition
- **Aesthetic Value**
  1. Landscaping quality

4 Results and Discussion

4.1 ARP Model

Considering the case of Chidambara Villas,

Constructed: **1904**

Extensions: **1920**

Estimated lifetime: 100 years (L_p)

Building age: 118 years (L_b)

O1 = 10% as there are no maintenance policies as such but it is being maintained on a regular basis as a utility

O2 = 0% as the heritage hotel is amidst a densely populated area and in close proximity to the nearby cities.

O3 = 0% as the layouts are flexible enough since most of them are being preserved without any modifications. Usage of services will happen for a long time.

O4 = 10% since the energy consumption levels are moderate. Only at times, renewable sources like solar energy are used.

O5 = 10% as the hotel is partially owned and also rented out to provide services for the tourists.

O6 = 0% since 90% of the mansion is preserved and only 10% is restored. Thus, the original design is followed reflecting the quality of the design.

Discount rate = **0.003**

Thus, the useful life of the building L_u = **74 years**

Effective useful life = 74 yrs. (El_u)

Effective building age = 118 yrs. (EL_b)

Effective physical life = 100 years (EL_p)

ARP decreasing formula is used to calculate the adaptive reuse potential since effective building age is equal to or more than its useful life.

$$(EL_b \geq EL_u)$$

The ARP value is **31.3**, which shows a moderate value for the adaptive reuse of the building. Hence, adaptive reuse is an apt choice to sustain the life of the building in this context.

4.2 Delphi Method

The questionnaire was circulated among a group of 50 people out of which 30 people responded. 15 people were architects; 10 were tourists, 2 were conservation architects, and the remaining 3 were transport planners and project managers. The questions that were circulated are as follows:

1. Has the mansion structurally improved? Is the original function being maintained?
2. Has the architectural style of the building been preserved? Are the new materials used in cohesion with the older ones? Rate the current interior quality of the heritage hotel.
3. Have the locations been taken care of? Is the setting of the building an apt one?
4. Have the transport facilities been taken care of? Are there separate means for vehicular and pedestrian access?
5. The Landscape features—Are there green breathing spaces around the mansion? Has the external space been legibly used?
6. After its conversion to a heritage hotel attracting a lot of tourists, has its economic value improved before (approx.)?
7. Have the cultural beliefs and traditional way of life of the Chettiar community been reflected in the heritage hotel?
8. Does the heritage hotel facilitate social interactions? Does it stand as an exhibition of different values?

9. Rate the aesthetic quality of the building in terms of its materials, façade works, and landscape.
10. Any other suggestions for the preservation or conservation of the heritage hotel for the sustenance of the life of the building even further?

From the survey, it is deduced that about 85% of the respondents have awarded low scores for the setting and landscape quality of the building. The building should coexist with the environment, and better usage of outdoor spaces, rather than just lush greens was suggested for the same.

For the adaptive reuse and the associated values to sustain even further, methods to discover new materials that would go with the physical structure without any compromises, creating more awareness amongst people about the culture and traditions associated with it through workshops, heritage walks, and experiencing the way of life were also suggested.

4.3 Discussion

The study shows the adaptive reuse of potential of Chidambara Villas. The significance of the research is the protection of cultural heritage in the Chettinad region. The reuse of existing structures also promotes sustainable development, rejuvenating the local communities and increasing tourism. Similar studies can assist conservation efforts contributing to the heritage, environment, and the economy of the region.

5 Conclusion

Determining the adaptive reuse potential before suggesting it as the best choice for the conservation of a structure becomes a must. The ARP model assesses the useful life of a historic building by considering the factors that influence obsolescence. As a result, the model can assist planners and policymakers in prioritizing the current inventory of buildings in terms of what are the most effective adaptive reuse projects, before valuable time has been invested in detailed analysis.

The Delphi method is best when different people's opinions and thoughts ought to be considered in order to make decisions. Thus, it provides a different analysis and various options for the same issue to decide. Since it is also anonymous, it avoids conflicts between experts and encourages creative participation. Both methods have their own advantages. Depending on the context of the situation, any one or even both methods can be used for a better determination of the adaptive reuse potential of any structure.

References

Australia ICOMOS (1979) The Australia ICOMOS Guidelines for the Conservation of Places of Cultural Significance ('Burra Charter'). http://australia.icomos.org/
Australia ICOMOS (1999) The Burra Charter" Electronic document. http://australia.icomos.org/publications/charters
Behance (no date) The revive Chettinad Project, Behance. https://www.behance.net/gallery/7803135/The-Revive-Chettinad-Project. Accessed: 23 May 2023
Chen F, Hong Y (2017) Evaluating the adaptive reuse potential of buildings in conservation areas Evaluating the adaptive reuse potential of buildings in conservation areas: the Grand Canal area in Hangzhou, China. Facilities Structural Survey Iss 354(18). https://doi.org/10.1108/F-10-2015-0077
Chen CS, Chiu YH, Tsai L (2018) Evaluating the adaptive reuse of historic buildings through multicriteria decision-making. Habitat Int 81.https://doi.org/10.1016/j.habitatint.2018.09.003
Chennai Nagarathar (No date) Chettinad Map. https://chennainagarathar.tripod.com/Chettinad map_Copy_1.html. Accessed: 23 May 2023
Dhakshinamoorthi J, Sasidhar K, Vedamuthu R (2014) Adaptive reuse and restoration of a Chettinadu mansion, Pudukottai, Tamil Nadu, India. In: SAHC2014—9th international conference on structural analysis of historical construction
Dokras U (2021) Ancient Tamil Economic Adventurers. CONSULTANT Human Resource and Administration
Hearn MF (1990) The architectural theory of Viollet-le-Duc readings and commentaries. Massachusetts Institute of Technology Press
Kiruthiga K, Thirumaran K (2017) Visual perception on the architectural elements of the built heritage of a historic temple town: a case study of Kumbakonam, India. Front Architectural Res 6(1). https://doi.org/10.1016/j.foar.2016.10.002
Lipe WD (1983) Value and meaning in cultural resources. National Archeological Database (NADB), tDAR (The Digital Archaeological Record)
Meenakshi M, Muthiah S, Ramaswamy V (2010) The Chettiar heritage. The Chettiar Heritage Press
Michell G, Ramamrutham B (2015) Mansions of Chettinad. Meenakshi Meyyappan and Graf media
Mohan K (2019) Change in spaces: adaptive reuse from residential to commercial. Vellore Institute of Technology
Natarajan S (2015) Chettinad architecture. Academia.edu
Patwardhan A (2018) Architecture: an heirloom in the context of Chettinad, India. ATINER's Conference Paper Series ARC2017-2393
RajivKumar S (2019) Analyzing the values in the built heritage of Chettinadu region, Tamil Nadu, India. Creative Space 7(1). https://doi.org/10.15415/cs.2019.71005
Rajmohan S (2016) Restoring lost Splendour—The Hindu. The Hindu. https://www.thehindu.com/
Riegl A (1996) The modern cult of monuments: its essence and its development. Historical and Philosophical Issues in the Conservation of Cultural Heritage. Getty Conservation Institute
Ruskin J (1849) The seven lamps of architecture. The Waverley Book Company
Subramanian TS (2018) Palaces of Chettinad. Frontline-The Hindu. https://frontline.thehindu.com/
UNESCO (New Delhi) (2010) Indian Heritage Passport Programme: on the Chettinad trail in Tamil Nadu; a concept paper. The Government of Tamil Nadu, Department of Tourism & Culture. https://unesdoc.unesco.org/ark:/48223/pf0000188788
UNESCO (World Heritage Convention) (2014) Chettinad, village clusters of the Tamil merchants. https://Whc.Unesco.Org/En/Tentativelists/5920/

Holistic Study, Analysis of Adaptive Reuse of Historic Structure

Vijayalaxmi J., Mothi Krishna, and Harshitha Sahu

Abstract Adaptive reuse is the modification of function in heritage structures without compromising on cultural, aesthetic, and traditional values. This study aims to evaluate the obsolescence of a historic building to determine the time for a change in its function. The Victoria Hall in Chennai is assessed for its qualitative and quantitative life. The Adaptive Reuse Potential (ARP) model is used to evaluate the physical life, useful life, and building's age to assess the potential for reuse. The results show that an Adaptive Reuse Potential (ARP) Score of 67.74% indicates the increased need for the adaptive reuse of Victoria Public Hall, Chennai, which will eventually result in sustenance as a heritage building for a longer time period. Findings indicate that the adaptive reuse technique can be carried out within the stipulated timeline for improved building life. This study is important to understand the conservation potential and process of adaptive reuse of historical buildings to improve the economic, social, and functional aspects.

Keywords Adaptive reuse potential (ARP) model · Historic buildings · Obsolescence criteria · SDG11

Abbreviation

ARP Adaptive Reuse Potential

Vijayalaxmi J.
School of Planning and Architecture, Vijayawada, India
e-mail: vijayalaxmij@spav.ac.in

M. Krishna (✉)
Space Matrix Design, Pvt. Ltd, Bengaluru, India
e-mail: mothikrishna64@gmail.com

H. Sahu
Architect, Organo Eco-Habitats, Hyderabad, India
e-mail: hsahu24@gmail.com

Vijayalaxmi J. (ed.), *Conservation of Built Heritage in India*, Advances in 21st Century Human Settlements, https://doi.org/10.1007/978-981-99-8335-3_10

1 Introduction

The need for the conservation of architectural heritage in urban areas has become very important as they are highly prone to succumb under the pressure of urban land economics. With growing years, heritage buildings face a lack of appreciation and care primarily due to the cost implications of their maintenance. Conservation of these structures through adaptive reuse is economically sustainable and preserves the cultural heritage. Adaptive reuse is a conservation technique wherein the heritage building is retained by putting it to a different function than what it was built for. By assessing the economic, social, and environmental aspects, the justification for the reuse of the heritage building can be established.

The adaptive reuse potential can be assessed for any building, prior to the adaptive reuse, with the help of an Adaptive Reuse Potential (ARP) model. The Adaptive Reuse Potential (ARP) model is used to assess and evaluate the life of a heritage building. It considers seven different obsolescence criteria to determine the physical life of the building.

The study by Doroz (2019) indicates that adaptive reuse can be carried out in structures that are vulnerable in terms of physical condition and structural integrity. The adaptive reuse technique includes recommendations and suggestions for adjusting such vulnerable structures to new requirements with a modern function and bringing a new life to them in the current environment (Cerreta et al. 2020). The author also studied Kielce, a city in Poland where actions were undertaken as part of the urban revitalization to enhance the functionality and aesthetics of the streets. The author highlights the need for the improvement of basic street elements and further analyzes the buildings subjected to adaptive reuse. This demonstrates the significance of the fundamental components leading to the larger developed structure.

A study by Conejos and Langston (2010) states that adaptive reuse is an emerging and significant design strategy that supports global climate protection and emissions reduction which is an important SDG 11 goal.

In another study by Yaldiz and Asatekin (2019), the author highlights the need to reuse the elements that reflect the heritage and cultural values of ancient monumental buildings. The author also emphasizes that new functions assigned to the monumental buildings should not affect their cultural value or original identity, and also it shouldn't alter the current maintenance policies. The author considers three major aspects such as spatial formation, the structure of the building, and conservation measures as major parts of the adaptive reuse of heritage buildings.

The study by Basha et al. (2017) discusses the correlation between adaptive reuse and social sustainability (the relationship between the building and the local community and its people) of the heritage structure. The author found that there are not enough studies being done on socially sustainable parameters (cultural, aesthetic, social, and human comfort). Hence, she took the case study Asia Heritage Row building located in Doraisami Street and Jalan Yap Ah Shak, Kuala Lumpur, and collected data through primary and secondary sources in the form of questionnaires and literature review, respectively, from the end users of Asia Heritage Row. The findings show that the adaptive reuse of the Asia Heritage Row building has significant value both culturally and socially. It preserves the heritage significance as a place of social interaction and ensures its sustainability.

From the above studies, it is evident that it is essential to consider adaptive reuse for heritage buildings to prevent them from becoming obsolete. Adaptive reuse can also ensure the sustainability of heritage buildings along with conserving allied buildings and street features while ensuring public participation. Building's physical life, age, and structural integrity along with maintenance policy are essential for adaptive reuse. Thus, this study aims to bridge the gap between these aforementioned components.

2 Study Area

2.1 Study Building—Details

Victoria Public Hall is an iconic heritage building, built in the late nineteenth century and it is located in Chennai (Fig. 2). The public hall has the characteristics of Indo-Saracenic Architecture and was designed by Robert Fellowes Chisholm, a British architect who pioneered in Indo-Saracenic Architecture in Chennai. The construction of the structure started in 1886 and continued for 4 years before its completion in 1890. The hall served the purpose of hosting important public and social events. Figure 1 shows the building in its completion.

The Public Hall extends in a land area of over 3.14 acres and the structure was built with three stories with the ground floor having a built-up area of 13,342 sq. ft.

Fig. 1 Victoria Hall—late nineteenth century (*Source* www.navrangindia.in, 2017)

Fig. 2 Victoria Hall—early twenty-first century (*Source* The Victoria Public Hall (1890), Chennai/en.wikipedia.org)

and the first floor is built with 12,541 sq. ft. The Public Hall consists of two halls with a seating capacity of 600 on both the first and second floors and the verandahs on the first floor were supported by the Slender Corinthian columns (Fig. 3), with a square tower that rises to three stories and with a pyramidal roof in the top with Islamic calligraphic inscriptions on it.

2.2 Why Victoria Hall

Since the construction of Victoria Hall in 1890, the first renovation work of the building took place in 1967, under the regime of the former Chief Minister Mr. C. N. Annadurai. However, after the successful operation of another 18 years, post renovation, the structure was subjected to legal disputes over the end of the lease period of 99 years. There were long-standing legal issues between Chennai Corporation and the Victoria Public Hall trust and it came to an end in 2009 with a verdict in the favor of Chennai Corporation, with the building being taken under the CMDA heritage structure Grade I Category.

Since then, officials of Chennai Corporation planned for the renovation and adaptive reuse of Victoria Hall, which has been closed for over 20–25 years, since 1985

Fig. 3 Victoria Hall—exterior elevation with windows supported by Slender Corinthian Columns (*Source* http://newindianexpress.com/, 2022)

due to lease issues. Chennai Corporation started the renovation work on April 1, 2009 (Fig. 4). However, the works were stalled due to various reasons such as lack of funds and construction of Chennai Metro Rails (CMR), which made the erection of scaffolding works for the renovation of Victoria Hall stop, as it affected CMRL works. The Chennai Corporation has proposed the plans of reusing the hall as a museum along with its original function of being a public hall for social events.

Fig. 4 Victoria Hall—under renovation (*Source* http://newindianexpress.com/, 2021)

Thus, in this study, we are going to conduct a detailed analysis of the Adaptive Reuse of Victoria Hall (museum) since its latest renovation in 2009, with the ARP model and come up with discussions on the basis of the ARP score, regarding the useful life of the building post renovation work.

3 Materials and Methods

3.1 ARP (Adaptive Reuse Potential) Model

The **ARP** model was developed by Langston (2008, 2012). The model aims to identify and rank adaptive reuse potential for the existing buildings and can therefore be described as an intervention strategy to ensure that collective social value is optimized and future redundancy is planned. The rating for the heritage properties and other buildings is done on the basis of conversion potential and timeline of reuse.

ARP model has the ability to undergo major changes in any building without losing its original features. It also helps in finding sustainable material that can be reused, before sending it to the landfill sites. The model can be used to assess the reuse potential of any building irrespective of location. ARP model uses the building's age, useful life, and physical life to assess the reuse potential. Along with the mentioned parameters, it also uses the obsolescence criteria for finding out the reuse potential.

ARP model uses the data from the above-mentioned parameters to give reuse potential values in percentage. With this percentage, one can assess various buildings in a region or locality and rank them to prioritize the buildings with high ARP percentages for immediate initiation for adaptive reuse. A higher ARP percentage indicates that the age of the building is close to its expected useful life.

The range of Adaptive Reuse Potential value (ARP Score) falls between 0 and 100%, where the higher the ARP value, the higher the potential, i.e., higher the chance or need for the change of use or function of the building. There is a set of formulae shown in Table 1 for the APR calculation.

Major terms used in ARP calculation are:

L_p—Physical life, calculated in years based on the questionnaire prepared across three different contexts, namely, environmental background, occupation details, and structural integrity, and filled by the experts and conservation architects.
L_u—Useful life, which can be calculated by equation one in Table 1.

Table 1 Formulae used in ARP model

S. no.	Formula	Description
1	$\mathbf{L_u} = \frac{\mathbf{L_p}}{\left(\mathbf{1+\sum_{i=1}^{7} O_i}\right)^{\mathbf{L_p}}}$	This formula is used to calculate useful life (L_u)
2	$\text{ARP} = \frac{100 - \frac{{EL_u}^{2*EL_b}}{100}}{EL_u}$	This formula is used to calculate ARP Score

$\mathbf{L_b}$_Building's age.

$\mathbf{EL_p}$ (Effective physical life), $\mathbf{EL_u}$ (Effective Useful life), and $\mathbf{EL_b}$ (Effective life of building/building's age) are obtained by multiplying $\mathbf{L_p}$, $\mathbf{L_u}$, and $\mathbf{L_b}$ with **(100/** $\mathbf{L_p}$**)** respectively.

$\mathbf{O_i}$_Summation of percentages of obsolescence criteria.

3.2 Obsolescence Criteria

The useful life of the building plays a major role in finding the ARP percentage. However, useful life is affected by seven obsolescence criteria, namely, physical, economic, social, functional, technological, legal, and political obsolescence.

Physical: Physical obsolescence mainly deals with the maintenance of the structure. The performance of the building depends upon the maintenance policy and its useful life tends to reduce if the building elements are poorly maintained.

Economic: Economic obsolescence majorly depends upon the location of the building. The structures found in the CBD (central business district), economic hubs, or market places were found to undergo less economic obsolescence whereas structures in remote areas were found to be economically obsolete.

Social: Social obsolescence is based on the relation between the building's function and its locality. If a structure depends upon external income or if the structure doesn't meet the function it is intended for, then the building is said to be obsolete on the social grounds.

Functional: Functional obsolescence is based on the extent of flexibility in the building's design. The extent of flexibility has an impact on the useful life of the building. Buildings with more flexibility in their design are found to be less functionally obsolete.

Technological: Technological obsolescence is based on the energy demand of the building. If a building is using high external energy for its occupant's comfort, then it has a high chance of technological obsolescence in the future.

Legal: Legal obsolescence deals with the quality of the building's design. If a structure is built with high standards, it is more likely to face less legal obsolescence in the future.

Political: Political obsolescence depends upon the level of interest of the public or local community in a particular building. If a structure has frequent political interference, it is more likely to undergo political obsolescence which eventually affects the useful life of the building.

All seven obsolescence criteria have a scale of 0–20%, where 0% indicates that the structure is efficiently safe while 20% indicates the structure is at high risk, based on each criterion. Also, the structure with intermediate effect will account for 5–10%, which is used in the useful life and ARP score calculation.

3.3 Literature Case Study

Application of the ARP Model to Lui Seng Chun:

Lui Seng Chun is located at 119, Lai Chi Kok Road in the Mong Kok district, one of the most densely populated areas of Hong Kong. The historic building was designed as a Chinese shophouse over four levels, exhibiting wide veranda (external corridors) on most facades (Fig. 5).

The building was constructed in the year 1932. Hence, the current building age is 90 years. The physical life is estimated to be 100 years.

- The maintenance was minimal for its entire life and has been left in a state of disrepair. So, a score of 20% has been assigned.
- The building receives a 0% reduction as it sits in a retail center in of the most densely populated areas of Hong Kong.
- The building being largely open design would attract a low churn cost for alterations. So, a reduction of 5% has been assigned.
- The building has a small narrow floor plan with high ceilings and large covered balconies all around. It relies on mechanical systems for occupancy. Thus, a value of 0% for technological obsolescence has been assigned.
- Although the structure functioned as a shopfront with a residence above, it has obtained income through retail or accommodation services. A 20% reduction is thus assigned.

Fig. 5 View of the Lui Seng Chun building

- Lui Seng Chun is sturdily built and is of high standard right from 1932 till today, although some structural rectification work is evident. A 5% reduction is assigned.

Using these data in the ARP model, the useful life is calculated as **61 years** and its adaptive reuse potential is **41.6%**.

4 Results and Discussion

4.1 ARP Model—Victoria Hall—a Heritage Building in Chennai

4.1.1 Physical Life Calculations

The physical life of the building deals with structural safety, that is, the time period for which the building can accommodate occupants safely without any threats. This physical life of the building can be found by analyzing three important components such as environmental component, structural integrity, and occupational aspect. All three aspects will be analyzed from primary data through a questionnaire survey by conservation architects in the field who have visited and studied the structure.

This physical life is expressed in terms of years on a scale being 0, 25, 50, 75, 100, 150, 200, 250, and with 300 years being the maximum. Figure 6 shows the primary response of people who have visited the site in person and based on their observation of questionnaires the physical life is calculated.

This graph shows the response of each individual under three categories, namely, Environmental Context, Occupational, and Structural integrity in terms of years in the y-axis. However, the total final physical life is cumulative of all three aspects.

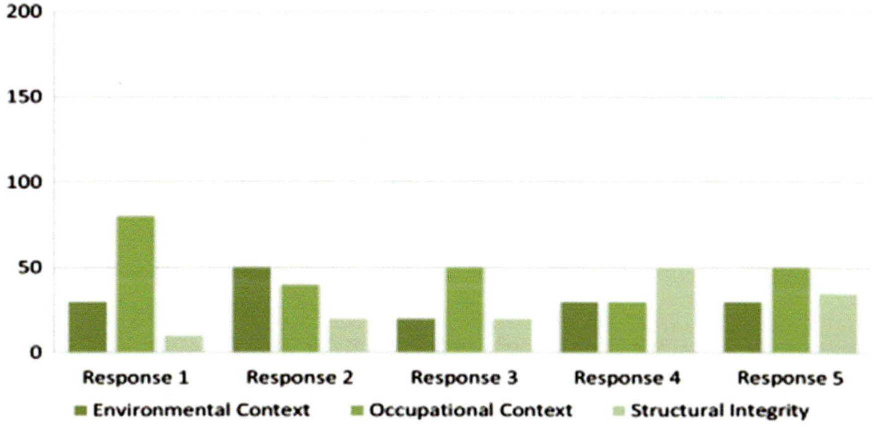

Fig. 6 Graphical representation of Questionnaire survey for physical life calculation

For example, Response 1 (30 yrs + 80 yrs + 10 yrs = 110 yrs) is rounded off to 100 years. Similarly, Response 3 had a breakup like (20 yrs + 50 yrs + 20 yrs = 90 yrs). Likewise, all five responses were found to be nearly 100 years.

Thus, the physical life of Victoria Hall is found to be 100 years, which means the building is structurally safe for the next 100 years since its first major renovation which took place in 1967.

$$\text{Physical Life; } L_P = 1967 + 100 = 2067$$

Till 2067 the building is structurally safe.

4.1.2 Useful Life Calculation

The useful life of any building can be defined as its habitable life that is fit for human use along with the structure serving the original purpose for which it is intended. Useful life (L_u) is calculated using the formula given in Eq. 1 of Table 1. For calculating the Useful life, one has to find out the cumulative obsolescence under the seven categories mentioned earlier.

The graph in Fig. 7 shows the obsolescence of each factor as a percentage with 0.0% being the lowest and 0.20% being the highest on the y-axis. This graph gives the cumulative obsolescence percentage of all the responses (individuals) along with the physical life of the building. These data are also based on the responses from the individuals who have visited the structure in person.

The summation of all the obsolescence criteria is ($\sum_{i=1}^{7} O_i = 0.18\% + 0.0\% + 0.07\% + 0.08\% + 0.15\% + 0.1\% + 0.06\% = 0.64\%$).

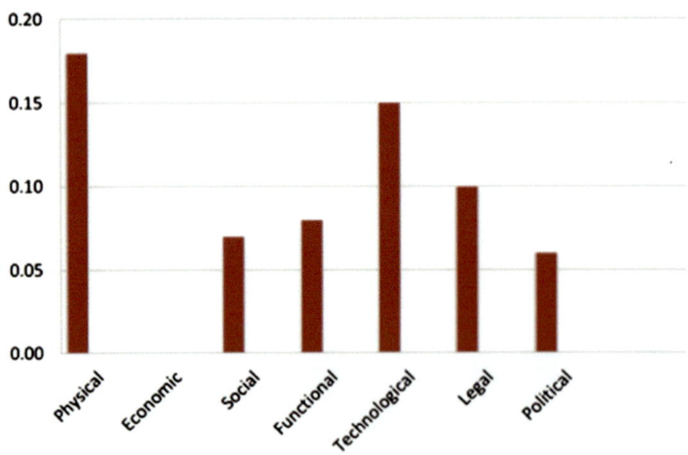

Fig. 7 Graphical representation of obsolescence criteria in percentage

$$\text{Useful life formula} = L_p / \left(1 + \sum_{i=1}^{7} O_i\right)^{Lp}$$

All the unknowns like physical life (L_p) and summation of all the obsolescence percentages are known. By substituting the values in the above formula, the useful life (L_u) is found to be 55 years.

The useful life of Victoria Hall is 55 years, Since its first renovation work in 1967.

1967+55= 2022.

4.1.3 ARP Score Calculation

From Eq. 2 of Table 1, the ARP score can be calculated. The unknowns like effective useful life (EL_u), effective building life (EL_b), and effective physical life (EL_p) are calculated by multiplying $\mathbf{L_u}$, $\mathbf{L_b}$, and $\mathbf{L_p}$ with 100/ L_p respectively.

$$\begin{aligned} \text{Effective Physical life } (EL_p) &= L_p * 100/L_p = 100 \text{ years} \\ \text{Effective Useful life } (EL_u) &= L_u * 100/L_p = 55 \text{ years} \\ \text{Effective Building life/Age } (EL_b) &= L_b * 100/L_p = 54 \text{ years} \end{aligned}$$

By substituting the above value in Eq. 2 of Table 1, the ARP score is 67.74%, which is high. The higher the ARP score higher the chances for adaptive reuse or the higher the need to change the functionality of the building. Thus, the findings can be summarized as.

5 Discussion

From Table 2, it can be seen that the building's age is taken as 42 years considering it till 2009. This is because as mentioned earlier, the Chennai Corporation is renovating the structure since 2009 and it is still ongoing. The ARP report (summary of findings) can serve as an additional supporting document for renovating and opting for adaptive reuse by including the design and function of the museum along with the existing structure. The remaining useful life of 13 years implies that the structure can function in the existing condition till 2022. However, after that, the building requires a change in functionality or adaptive reuse in order to sustain on economic and social grounds. Thus, by using the ARP tool, one can calculate the useful life, physical life, and remaining useful life of a historic building, which helps to frame an appropriate timeline for adaptive reuse and timely renovation.

Table 2 Summary of findings and their corresponding data—Victoria Hall, Chennai

Summary	Details
Original/actual construction of the building	1890
Last major renovation of the building	1967
The most recent renovation of the building	2009 (ongoing)
Physical life of the building (L_p)	100 years
Useful life of the building (L_u)	55 years
Age of the building (L_b) (Till 2009)	42 years
ARP rate (%)	**67.74%**
Remaining years of useful life	13 years (from 2009)

6 Conclusion

In this study, with the help of the ARP tool, the study of a building is conducted to prioritize them based on the need for adaptive reuse with reference to their ARP score. Conservation of heritage or historic buildings should be carried out at an appropriate time and within the country's laws, schemes, plans, and policies. It is also aimed under SDG11 (Sustainable Development Goal) under target 11.4 to enhance efforts for protection of the cultural and natural heritage under sustainable cities and communities. Renovation and Restoration of historic buildings should be carried out by trained professionals and any mismanagement in labor could lead to damage to the structure which may lead to loss of original identity. Another important factor is the lack of funds, which results in reduced sustenance of historic buildings in the long run. Based on the obsolescence criteria, one can learn and take initiatives to reduce that obsolescence, so that adaptive reuse has much scope in the future and the cost spent on the same activity repeatedly can be reduced.

References

Basha B, Abdullah F, Soomro AR (2017) Viability of adaptive reuse with social approach: case study of Asia heritage row building, Kuala Lumpur. Adv Sci Lett 23(7):6179–6183. https://doi.org/10.1166/asl.2017.9231

Cerreta M, Elefante A, La Rocca L (2020) A creative living lab for the adaptive reuse of the Morticelli church: the SSMOLL project. Sustainability 2020(12):10561. https://doi.org/10.3390/su12241 0561

Conejos S, Langston CA (2010) Designing for future building adaptive reuse using adaptSTAR. In: International conference on sustainable urbanization. The ICSU, Hong Kong

Doroz M (2019) Revitalization of small towns and the adaptive reuse of its cultural heritage. In: IOP conference series: materials science and engineering, vol 471. https://doi.org/10.1088/1757-899X/471/5/052006

https://www.newindianexpress.com/cities/chennai/2021/jul/16/victoria-hall-to-be-converted-into-rolling-museum-2330811.html, Published 16th July 2021 06:04 AM

https://www.newindianexpress.com/cities/chennai/2022/jun/06/chennais-victoria-hall-renovated-to-host-public-meetings-2462263.html published on 6th June 2022, read on 9th Oct 2023

Langston CA (2008) The sustainability implications of building adaptive reuse. Paper presented at the Chinese research institute of construction management (CRIOCM) international symposium, Beijing, China

Langston C (2012) Validation of the adaptive reuse potential (ARP) model using iconCUR. Facilities 30(3–4):105–123. https://doi.org/10.1108/02632771211202824

http://www.navrangindia.in/2017/01/victoria-public-hall-colonial-madras.html, Published on 5th Jan 2017, read on 9th Oct 2023

The Victoria Public Hall (1890) Chennai/en.wikipedia.org read on 9th Oct 2023

Yaldiz E, Asatekin NG (2019) Definition of a new design process for the reuse of monumental buildings. In: Proceedings of the central Europe towards sustainable building 2013

MIX
Papier aus verantwortungsvollen Quellen
Paper from responsible sources
FSC® C105338

If you have any concerns about our products,
you can contact us on
ProductSafety@springernature.com

In case Publisher is established outside the EU,
the EU authorized representative is:
Springer Nature Customer Service Center GmbH
Europaplatz 3, 69115 Heidelberg, Germany

Printed by Libri Plureos GmbH
in Hamburg, Germany